Blind Landings

Blind Landings

*Low-Visibility Operations in American
Aviation, 1918–1958*

ERIK M. CONWAY

The Johns Hopkins University Press
Baltimore

© 2006 The Johns Hopkins University Press
All rights reserved. Published 2006
Printed in the United States of America on acid-free paper
2 4 6 8 9 7 5 3 1

The Johns Hopkins University Press
2715 North Charles Street
Baltimore, Maryland 21218-4363
www.press.jhu.edu

Library of Congress Cataloging-in-Publication Data
Conway, Erik M., 1965–
Blind landings : low-visibility operations in American aviation, 1918–1958 /
Erik M. Conway.
p. cm.
Includes bibliographical references and index.
ISBN 0-8018-8449-7 (hardcover : alk. paper)
1. Instrument flying—United States—History—20th Century. 2. Airplanes—
Landing—United States—History—20th Century. 3. Landing aids (Aeronautics)—
United States—History—20th Century. I. Title.
TL711.B6C66 2006
629.132′521409730904—dc22   2006003795

A catalog record for this book is available from the British Library.

To my parents, Richard and Barrie-Lynn Conway

Contents

Acknowledgments

A dozen years ago, I was afloat in the vast tracklessness of the Pacific Ocean when I received an e-mail from John Beatty welcoming me to the University of Minnesota's Program in the History of Science and Technology. "Welcoming" proved an understatement, and I want to thank John, Sally Gregory Kohlstedt, Arthur Norberg, Robert Seidel, Alan Shapiro, Roger Stuewer, and John Euler for fostering an atmosphere of collegiality amid the horribly cold Minnesota winters. The friendship and camaraderie offered there made a sometimes-difficult transition from navy officer to (hopeful) scholar the most pleasurable years of my life.

The idea for this project came from a conversation with another of Minnesota's graduates, Robert Ferguson. I am certain that this is far from what he expected it to be, but I hope that it is still satisfactory. My friends Eileen and Andrew Weingram provided a comfortable home away from home on too rare vacations. And I cannot thank Barton C. Hacker and Margaret Vining enough for the many research visits they have hosted at their Clio house in Washington, DC, over the years.

As all historians do, I have accumulated debts to librarians and archivists at various institutions. At the New England branch of the National Archives, Helen Engle aided this project enormously through her knowledge of the MIT Radiation Lab papers and her work-in-progress finding aid, more than 600 pages long. Patricia White at the Stanford University Archives guided me to important files on klystron development and the Sperry Gyroscope Company's involvement in blind landing research. Mary Pavlovich, librarian at the National Air and Space Museum, was unfailing in her attempts to procure obscure materials via interlibrary loan. Marjorie McNinch at the Hagley Museum and Library provided access to the Aircraft Owners and Pilots Association records and additional Sperry Gyroscope Corporation records, and she arranged permission for use of Hagley images in this book. Kate Igoe at the National Air and Space Museum provided access to many images used in this book, and I am grateful for her efforts to identify and secure permission for several rare photographs.

Completing this project required access to records still held by operating agencies. Mr. V. Pattanyak, at the International Civil Aviation Organization's headquarters in Montreal, kindly provided access to an extensive collection of documents related to international technical negotiations. The records officer at the U.S. Federal Aviation Administration, Lisa McGlasson, provided access to many documents still in the agency's possession, while FAA's now retired historian, Ned Preston, provided key documents from its few remaining historical files. Finally, Hal Moses at the Radio Technical Commission on Aeronautics allowed me to use the organization's collection of 1940's documents. Without the assistance of these agencies and their representatives, this history would be poorer.

The assistance of survivors of the World War II–era U.S. Eighth Air Force was also invaluable. More than thirty veterans responded to an advertisement I placed in the Eighth Air Force Historical Society's newsletter to tell me about their experiences flying and landing in the European theater. Collectively, they provided an insight into wartime problems that I could not have gotten from the published literature. I cannot thank them all here, but Dr. John A. Clark, Col. Bruce Edwards, Col. Francis L. Grable, Charles W. Halpen, J. W. Howland, Joseph E. Manos, William D. Reeder, and Bill Rose took the time to write detailed letters about their experiences. Dan Raymond and George Reynolds sent me articles that had appeared in various other veterans' newsletters that I would never have found, and Chuck Derr sent me a self-published book on his wartime experiences. Most extensive of all has been the correspondence of Col. Robert W. Vincent. Thank you all for your sacrifices.

The National Air and Space Museum supported this project through a predoctoral Guggenheim Fellowship, freeing me from the usual demands of graduate student teaching and allowing access to the vast archival collections there. The U.S. Army Signal Corps alone collected more than 13,000 pages on blind landing work between 1933 and 1939, and the Air Corps, Navy Bureau of Aeronautics, and Civil Aeronautics Authority each accumulated similar volumes of material. Residence in the Washington area made this project feasible. The National Science Foundation, via dissertation improvement grant NSF/SBR-9729517, supported research conducted outside Washington. I largely completed the revisions to this manuscript while a contract historian at the National Aeronautics and Space Administration's Langley Research Center in Hampton, Virginia.

I am indebted to my advisors, Arthur Norberg at Minnesota and Michael Neufeld at the National Air and Space Museum, and to Sally Gregory Kohlstedt at Minnesota, for their careful readings of and comments on earlier versions of this manuscript. Conversations with my friends and colleagues Deborah G.

Douglas and Stuart G. McCook were important to my thinking about airfields and landing aids. Their thoughtful comments have improved this book immeasurably, yet all errors, whether of fact or interpretation, are mine alone.

Finally, I thank the National Air and Space Museum, the Hagley Museum and Library, Unisys Corporation, and the Aircraft Owners and Pilots Association for permission to reproduce images contained in this work, and to the Johns Hopkins University Press for permission to reproduce much of chapter 7 from an earlier article.

Abbreviations

A-1	a U.S. Army Air Corps landing aid, also known as the "Hegenberger system"
AAF	Army Air Forces
ALPA	Air Line Pilots Association
AN/MPN-1	a World War II–era GCA system
AOPA	Aircraft Owners and Pilots Association
ARINC	Aeronautical Radio, Inc.
ARL	Aircraft Radio Laboratory (Wright Field, Dayton, Ohio)
ATA	Air Transport Association
BABS	blind approach beacon system
CAA	Civil Aeronautics Authority (until 1940); Civil Aeronautics Administration (after 1940)
FCC	Federal Communications Commission
FIDO	a thermal fog dispersion device
GCA	ground-controlled approach
IATCB	Interagency Air Traffic Control Board
ILS	instrument landing system
ITRM/ITD	International Telephone and Radio Manufacture; International Telephone Development Company (same company as the Federal Telephone Development Corporation)
NAS	National Academy of Sciences
NBS	U.S. National Bureau of Standards
PGP	pulsed glide path
PPI	planned position indicator
RAE	Royal Aircraft Establishment
RTCA	Radio Technical Commission on Aeronautics
SBA	standard beam approach system
SCS-51	U.S. Army version of the instrument landing system
UAL	United Air Lines

UHF   ultra-high frequency (after 1944; frequencies between 3 MHz and
          30 MHz)
USSTAF   U.S. Strategic and Tactical Air Forces
VHF   very high frequency (after 1944; frequencies between 30 MHz and
          3 GHz)
YB   a U.S. Navy version of ILS

Introduction

Pilot and writer Antoine de Saint-Exupéry, in his classic novel *Night Flight,* told of a French mail pilot and his radio operator who, surrounded by storms while flying between Patagonia and Buenos Aires, were unable to find a place to land. The pilot's supervisor, driven by the need to conquer the distance between isolated cities, had ignored the worsening weather and insisted that the mail go through on time. Saint-Exupéry did not recount their deaths; instead, the two men simply vanished from the narrative.[1] Their fictional deaths, like the real deaths that occurred all too frequently in the first thirty years of the aerial enterprise, were counted as merely two more casualties in the struggle to bring about aviation's great future.

Saint-Exupéry's novel reflected Charles Lindbergh's own experience. Lindbergh was flying for a mail contractor one night on the St. Louis–Chicago run in 1926 and became lost in a dense fog. When he was nearly out of fuel, he descended as close to the ground as he dared given the unknown terrain and dropped a parachute flare that he carried for such occasions. He hoped that the flare would illuminate the ground, permitting him to land in a convenient field, but it did not. Using the last of his fuel to climb as high as he could, he jumped out, relying on his parachute to save his life. The next day, he rescued the gasoline-soaked mail from the plane's wreckage, which a farmer had found a couple of hundred yards from his house.[2] This happened to Lindbergh twice that year.

His experience was not unusual. Virtually all mail pilots in the 1920s had to make forced landings, using holes in the clouds to try to find any field big enough to land their aircraft in to circumvent a crash. A report generated by the Post

Office's Air Mail Service in 1925 detailed the magnitude of the problem. Between July 1924 and July 1925, the mail service's pilots made 750 forced landings, 554 of which (77 percent) were caused by the weather.[3] That report's recommendations set the next decade's research agenda for the Army Air Service, the airlines created by privatization of the air mail in 1925, the regulatory authority created the following year, and the National Bureau of Standards' radio laboratory.

Before 1940, researchers called it the "blind" or "fog" landing problem. Inventors, engineers, and scientists of diverse backgrounds proposed literally hundreds of technological solutions. These ranged from the suspension of lit balloons as markers above airfields to the use of high-powered X-rays as guidance beams. The balloon method was popular with British researchers in the 1920s; happily, the U.S. Army Signal Corps checked with radiation specialists before seriously exploring the X-ray proposal. Also popular with inventors were systems based on the perceived ability of infrared light to penetrate clouds and fog, systems using underground cables (and some with the cables held above the surface on poles), acoustic systems, and radio beam systems.

The sheer diversity of proposed solutions to the blind landing problem suggests how large it loomed. Most proposals came from private individuals who submitted them to the Army Air or Signal Corps in the hope that one of the agencies would find their idea worthy and develop it. Several proposals received significant attention from the Corps' investigators, but few were actually developed into functional prototypes. This book will explore some of these prototype systems. I have not been able to locate documents relating to every prototype, including one that gave important service during World War II. These historical orphans, whose existence is known but whose development I cannot document, will deserve mention in later chapters, but this book is built around documented development efforts. That means, in essence, that it is a history of government-funded development projects, primarily in the United States but also, to a lesser extent, in Western Europe. The majority of these projects followed directly from the recommendations made in the Post Office's reports, and hence we can trace much of the infrastructural development of commercial and military aviation to the Post Office's airmail experiment in the 1920s.

Not all blind landing work stemmed from the airmail experiment. World War II produced important techniques and technologies as well. The vast air war the Royal Air Force and the Army Air Forces pursued in Europe relied heavily on the technologies devised to solve the airmail carriers' problems while inspiring many more, some of which were made available after the war to civil aviation interests. The infrastructure necessary to support all-weather flying in the United States is thus a product of what a more conspiracy-minded scholar might call a "military-postal-industrial complex" that was joined in the late 1930s by two universities, Stanford and MIT.[4]

The goal that the Air Corps, the Post Office, and, after airmail privatization, the airlines shared was all-weather operations. They, the privately financed Guggenheim Foundation for the promotion of Aeronautics, and the National Bureau of Standards all exchanged knowledge and equipment to reach that goal. While the United States Navy also pursued blind landings, it kept its research secret and hence had little impact. The same is true of French work. During the 1920s, France pursued a much different set of technologies in secret, permitting only the barest of descriptions to be published and thus having no influence at all in the United States. Germany, in contrast, openly adapted American research after 1932, when it was largely freed from Versailles Treaty constraints. After World War II, U.S. technologies were adopted as the world standard, eliminating the diversity of techniques that marked prewar aviation.

Histories of aviation are often written as quest narratives, and the extant histories of blind landing research are of this type. James Hansen has criticized historians of "going native" when studying aviation and reinforcing the master narrative of linear development and untrammeled success. Failures are thus typically forgotten. The blind landing story has been told in this way in the past.[5] But such an approach disguises a far more interesting story. While blind landings have been achieved, they have not become routine. Instead, the tremendous effort that went into solving the blind landing problem produced an infrastructure capable of supporting almost all-weather operations. Yet the goal, completely blind landings, remained elusive at the end of aviation's first century despite substantial advances in precision operations. This book thus will investigate a seeming paradox: while scientists and engineers developed systems increasingly able to provide reliable blind landings, pilots and regulators became less and less convinced that routine blind landings were either feasible or even a good idea. The result was that equipment intended to produce blind landings was installed worldwide but was used for a lesser purpose.

This book is both a history of the technologies of blind landing and a history of the idea of blind landings. It will trace the technologies of blind landing—at least some of them—through the cycle of invention, testing, improvement, negotiation, and adoption, while also following the gradual collapse of the concept of "blind landings." Through following these two threads, this book seeks to explain why the quest for blind landings failed.

When I first began looking into the history of aircraft landing aids, to use the post–World War II term, I was struck by the use of the terms "blind flying" and especially "blind landing" in the popular aviation magazines and newspaper articles of the interwar period. Living at the tag end of the twentieth century, and jaded with the knowledge that blind landing was simply not done, the idea of attempting it struck me as both dangerous and silly. The safe thing to do if bad weather is predicted at your destination or along your chosen route is stay home, as generations of prospective pilots have been taught in ground school. To facilitate that obvious bit of wisdom, the first piece of national aviation infrastructure built in the United States was an aviation weather network, constructed during the middle 1920s.[6] Yet something drove fliers like Saint-Exupéry and Lindbergh to ignore wisdom and challenge the weather anyway. Understanding why they did so is our first task.

Aviation was the first form of transportation to seek perfect all-weather operations, at least in any organized way. Automotive pioneers made no attempt to make cars and trucks capable of blind driving, no sane sea captain sailed into port in a fog—that's what anchors are for—and trains quite literally stopped in their tracks when weather became so bad that the engineer could no longer see. These other forms of transportation had one advantage over airplanes in dealing with the weather in that they could stop and wait it out. Once airborne, an airplane must always be in forward motion. Circling to wait out the weather was an option for aircraft, but only within the limits imposed by fuel, as Lindbergh's story suggests. Hence one could argue that safety motivated the development of blind landing systems, but that argument runs afoul of my earlier safety point: if one's primary concern is flight safety, one stays firmly on the ground. Since safety issues could be resolved by not flying, safety was not the sole motivation behind the development of blind landing systems.

Saint-Exupéry's novel suggests both of the primary reasons aviation pioneers sought to achieve blind landings: profit and progress. The profit motive is fairly obvious. Aircraft that sit on the ground produce losses, not profits. One subtext of Saint-Exupéry's story is the pressure to perform the airmail supervisor placed on his men, both pilots and mechanics. In the United States the Post Office was the original source of the drive to beat the weather. While the Post Office itself was not a profit-seeking enterprise, its leaders intended the airmail service to become profitable so that it could be commercialized, thus forming an economic basis for commercial air service on a national scale. Bad weather interfered with the potential for profitability. Successful commercialization meant that the weather's effects on aviation had to be mitigated through some technological means. While the weather also interfered with trains, which the airmail pioneers considered their primary competition, the interference was much less. This was because visibility at ground level is almost never completely absent, while solid overcast at a thousand feet or so is quite common.

Perhaps more important to the profit motive, however, was aviation's "dirty little secret": the airplane was not, in a practical sense, faster than trains were when airmail service began. While airplanes might cruise at 100 mph and the iron horse steamed along at only 40 mph, the train still won because aircraft had to stop at night and for bad weather. Trains could travel round the clock, while a wintertime airplane flight had only eight hours of daylight and spent about an hour of that on the ground getting fuel. Worse, mail pilots often found that they might be able to land in worsening weather but not take off again, introducing another sort of delay. Realization that the airplane was actually slower than the railroad was quick in coming. The Post Office tried several remedies beginning in 1920, but the most successful was a string of beacons, in reality extremely bright searchlights on rotating platforms, that it erected during 1923–24 to construct a transcontinental airway able to support round the clock flying. Mail pilots flew from searchlight to searchlight, and most of the original group of mail pilots died doing it. Yet the searchlights fixed the "night flying" problem well enough so that the airmail could beat the trains in good weather conditions.[7]

Once the Post Office had proven night flying possible and profitable, it privatized the airmail, contracting with groups of entrepreneurs to provide airmail service over the Post Office's old routes. These entrepreneurs founded the airlines that dominated U.S. commercial aviation through deregulation in 1978. The Post Office provided a deliberately generous subsidy to the airmail companies to ensure the service thrived. The Post Office's leaders also chose and assigned routes to expand service and prevent excessive competition. Finally, it structured the mail contracts to encourage airmail companies to carry passengers. Passenger service, in turn, demanded that the airlines achieve a much greater consistency of operation than the Post Office did. The mail cared very little if it arrived a few hours late due to bad weather or wound up in a different city than expected. Passengers accustomed to scheduled rail service were much less forgiving. Because the Post Office's, and airlines', goal was the creation of economically self-supporting scheduled passenger service, routine blind landing assumed substantial economic potential.

But the dependence of the air carriers on government subsidies (which continued in the United States into the 1950s) raises two other questions. What motivated governments to create these businesses in the first place and, in the second place, to keep them afloat financially? There are several parts to that question, and they all revolve around the western notion of "progress," especially the technological kind.

Three historians have discussed the modernist notion of "progress" in the context of aviation, from very different approaches. The first of these is Robert Wohl, who has examined western society's response to the airplane before 1918. He shows that Americans and Europeans responded to the airplane with enormous emotion once it had, beyond a doubt, been proven true by the Wright brothers' highly publicized demonstrations in 1908. Wohl calls the enthusiasm that the Wrights' efforts generated a "passion for wings." That passion inspired artists, architects, poets, and writers to present the airplane as a cultural symbol, while other young men, and a few young women, devoted their lives to promoting the new technology through air shows, races, and attempts to set speed, distance, or altitude records. Some people gave up otherwise successful careers to pursue the airplane. Briton T. E. Lawrence, who gained fame as the leader of an Arab uprising against the Ottoman Empire during World War I, gave up his officers' commission and the life of celebrity his exploits had earned him to become a private in the Royal Air Force, believing that "the air is the only first-class thing our generation has to do."[8]

The airplane's ability to break free from the earth, if only briefly, enabled it to become the vessel for long-held hopes for human betterment. Historian Joseph Corn calls the belief people like T. E. Lawrence held the "winged gospel": the airplane would improve the human condition when intercontinental flight permitted the free admixture of peoples and ideas. This, these true believers fervently hoped, would eliminate the misunderstandings they perceived as the causes of war. Corn writes that Americans "viewed mechanical flight as portending a wondrous era of peace and harmony, of culture and prosperity."[9] The airplane's ability to evoke such utopian imagery, he argues, descended in part from Christianity. Christian religious symbols, from the Star of David through the angelic hosts to the ascension of Christ, linked Heaven directly to flight. The flying machine, by its very nature, drew on these ancient symbols to stimulate a utopian vision of the future. That utopian aerial future would come about simply through the operation of unimpeded progress.

The religious symbolism of flight also reinforced another traditional thread in American culture, technological utopianism.[10] Corn traces this thread partly to the rise of evangelical Christianity and partly to the sheer pace of technological change in the nineteenth century. In a single lifetime, the development of large-scale production techniques and the deployment of mass transportation technologies of steamboat, steamship, and railroad had transformed American life. One could not live through such an age without developing an implicit recognition of the power of these new technologies to alter old patterns of life. Those transformations, in turn, fed upon the Enlightenment notion of progress and its seeming promise to deliver social improvement concomitant with the advance of technology.

Corn shows that this vision of "aerial progress" had strong democratic overtones. Belief that the airplane was a force for the spread of democracy ran deep enough that, before 1940, the federal government attempted to develop an airplane for the "everyman." There was wide public support for the goal of an "airplane in every garage."[11] In part, this was a reaction to the hated railroad trusts, since personal airplanes could allow the public to simply bypass them. It was also an extension of the public's response to Henry Ford's notion of personal automobility. Ford's own 1925 entry into airplane manufacturing was greeted with enormous enthusiasm by a public that believed he could make airplanes available to the masses the same way he had with the car. He failed, but others kept trying through the late 1940s, when the public's enthusiasm for the winged gospel finally faded. The dream was kept alive, however, by private fliers' clubs and lobby organizations, which will be important later in this book.

Yet there was more than one possible "aerial future" implicit in the winged gospel's vision. While Corn focuses on the mainstream version, he briefly acknowledges a minority thread within the vision: the airplane as mass transit. This was the vision held by Otto Praeger, the assistant postmaster general who built the initial airmail service; by Herbert Hoover, who as secretary of commerce helped produce the initial air carriers; by entrepreneurs like Great War ace Eddie Rickenbacker, who headed Eastern Airlines; by the near-legendary William Boeing, whose two air-related startups, Boeing Air Transport and Boeing Aircraft Company, eventually became the two largest businesses of their type in the world (United Airlines and The Boeing Company, respectively); and by investors like Henry Ford, William A. Rockefeller, and Cornelius Vanderbilt Whitney.[12] The airplane-in-every-garage was not at all what they perceived in the airplane. Instead, they imagined a ticket counter in every drugstore. This commercial vision of the airplane-as-mass-transit was much more important to the development of blind landing systems than was the more populist democratic vision. The drive to maintain reliable, scheduled service provided the direct impetus for many of the innovations this book documents. It was the commercial version of aerial progress that motivated government support and regulation of aviation in the early years, too, most clearly reflected in the Post Office's involvement, in the placement of aviation regulation in the Department of Commerce, and in the foundational legislation's very title: the Air Commerce Act.

Missing from Corn's book entirely is a third thread that was an inherent part of the aerial future: the airplane-as-ultimate-weapon. This was the dark side of aviation, recognized much more strongly in Europe than in America. Frenchman Jules Verne recognized the potential well before the Wrights actually flew, and British and Italian strategic theorists generated the doctrine known as "strategic bombing" during and after World War I. Yet plenty of Americans recognized, and approved of, this version of "aerial progress." Most famous among them was Brigadier General William Mitchell, whom the U.S. Army eventually court-martialed for publicly criticizing army aviation policy. Many historians of military aviation have documented Mitchell's public promotion of the airplane-as-weapon and his quest for a military air arm equal in stature (if not superior) to the army and navy. Seldom explicitly acknowledged was that Mitchell's proposed Department of Aviation would have placed *all* aviation under military control, a very different vision of aviation's future and one that General Henry "Hap" Arnold tried again to achieve in 1942 by working to militarize the Civil Aeronautics Authority.[13] Civil pilots and the airlines were vehemently opposed to military rule, although it was the navy's refusal to release the many naval reservists within Civil Aeronautics Authority to the army that directly blocked Arnold's attempt at domination. Mitchell's and Arnold's military vision of the aerial future was thus diametrically opposed to the ideal of democratic aviation because it envisioned centralized control over all aviation activities in the United States.

These three visions of aviation's future mostly manifested themselves in conflict over infrastructure. While one could design individual aircraft for specific purposes, the enthusiasts of the air all agreed that only one infrastructure to support aviation should be built. The selection of technologies for that infrastructure became the most hotly contested part of aviation development because advocates promoted technologies that "fit" their vision of the future. The plethora of blind landing system designs created during the interwar period provided wide latitude for this politicized process of selection to operate. On some occasions, the process of resolving disputes over blind landing system selection took the form of negotiations. In others, disputants resorted to congressional politics to get their way. Differing visions of aerial progress thus directly affected the development and deployment of blind landing systems.

In a recent book, historian Eric Schatzberg has investigated the role played by belief in what he called "the progress ideology of metal." He argues that belief in this ideology, which promoted metal as the "modern material" in contrast to old-fashioned wood, caused aircraft manufacturers, regulators, and purchasers to demand all-metal aircraft after the Great War. They did this despite what he calls the technological indeterminacy of the situation: researchers were never able to demonstrate convincingly that either material was superior in terms of strength, maintainability, or ease of manufacture. Schatzberg contends that this inability to prove objective superiority left a "social space" in which ideology could operate to influence the research agenda.[14] A similar analysis can be applied to the case of blind landing systems: several systems worked about equally well. None could be proven superior, and that lack of objectively demonstrated superiority allowed the three competing visions of aviation's future to influence the selection of blind landing systems.

Besides ideological motivations, there were several other major influences on the development of blind landing systems. "Nature" gave inventors, engineers, scientists, and entrepreneurs no end of trouble, and early hopes for a quick solution to the blind landing problem collapsed as the magnitude of nature's impact on the operation of these systems became clear. Both what William Cronon has called "first nature," the natural environment, and "second nature," the human-built environment, caused problems in roughly equal measure.[15] It would be too much to claim that these inventors negotiated with nature to achieve success, but they gradually learned that a rather intimate understanding of the effects of both regional and local operating environments, and a close tailoring of both the equipment design and specific, local installations to those conditions were necessary to achieve even a possibility of success. Systems that could not be tailored to local conditions were early casualties of the selection process. Even the mature, more flexible landing system adopted worldwide after World War II could not be tailored to all existing localities, and a great many airfields worldwide therefore cannot use it. While progress ideology was important as a motivator for technological change, "nature" influenced the technologies at a very detailed level.

So too did that amorphous, hotly debated thing called "human nature." I argue that the blind landing quest's failure was not an engineering failure but a human one. Scientists and engineers achieved technological systems capable of blind landings, a reality evident to anyone familiar with "smart weapons" and robot aircraft, both of which were in use by the waning days of World War II. If one can make a bomb find its own way to a target the size of a ship, one can certainly make an airplane find its way to a much larger runway. The first transatlantic flight of a fully robotized aircraft duly occurred in 1948.[16] Yet by this time the very idea of blind landings had vanished from the literature, replaced on the orders of

the armed services and the Civil Aeronautics Authority by a new term, "instrument approach." Flying and landing an aircraft "blind," using only the instruments provided in the aircraft, was ultimately an act of technological faith. The
evangelists of blind landing discovered to their dismay that average pilots and regulators had come to believe that no machine was trustworthy enough to actually
land an airplane without visual verification by the human crew. Hence the final
thread in this history is a social one. Pilots were the arbiters of technological
progress in aviation: technologies that did not earn their faith vanished; those
technologies that succeeded are still with us. Faith was neither automatic nor absolute. Pilots had to be convinced to trust their equipment enough to fly blind.
Most never achieved that level of trust, however. Faith had its limits, and we must
therefore examine how pilots, not just the famous great pilots but ordinary fliers
as well, responded to these technologies to get at those limits. This also means
we must examine the development of procedures for the use of blind flying and
landing technologies, because building reliable procedures and training pilots to
use them were key elements in the production of pilot faith.

By the end of World War II, the majority of the technologies of the modern air
traffic system had been developed in the course of efforts to produce all-weather
operations. Yet individual technologies were insufficient to solve the weather
problem. What no one foresaw before the war was the need for integration.
Wartime operations made that clear to aviation leaders, who began to make
progress toward integrated airport approach and landing control in the late
1940s. The effort stalled out as the nation returned to prewar levels of budget
stringency, but the immediate postwar effort included all of the prewar developments, radar derived from wartime programs, and, in a complete rejection of
blind landing, runway lighting. With the exception of digital computers, this suite
of technologies was the basis of the modern air traffic system, deployed nationwide after a series of catastrophic accidents created renewed political pressure for
reform.[17] While the construction of modern air traffic control is beyond the scope
of this work, its roots are in this immediate postwar drive for integration.

The book begins with an examination the development of blind, later called
"instrument," flying, to argue that the development of technique was as important as the development of technologies for flying blind. In Chapter 2, I explore
the relationship between airfield design and blind landing systems, to argue that
the conversion of paved runways in the United States during the late 1920s rendered one possible blind landing system unusable. The early history of the current "instrument landing system" is the subject of Chapter 3, which argues that
the instability of early versions helped delay the system's adoption while forcing

engineers to innovate to stabilize them. Chapter 4 examines one potential solution to the instability problem, microwaves, pursued jointly by Stanford, MIT, the Army Signal Corps, the Civil Aeronautics Authority, and the Sperry Gyroscope Company, in the context of disputes and negotiations between government agencies and airlines over whether to adopt improved versions of the original National Bureau of Standards system as the U.S. standard or the microwave system.

World War II serves as a major division in the book. The European air war marked the first attempt to use air power in all weather conditions, and the Allied forces quickly found that they needed a blind landing system—or several. Chapters 5 and 6 each examine the development and deployment of one such system, one using ultra-high frequencies (UHF), which eliminated the microwave system that had appeared so promising in 1939, and a radar based system that became famous as "ground controlled approach." Chapter 7 traces the postwar fight between supporters of these two systems over which to adopt as the national standard, to argue that the struggle represented a conflict between two very different visions of aviation's future. Chapter 8 examines the transformation of the two systems into an almost-all-weather system of landing aids, marking the final acceptance that routine blind landings were simply not humanly feasible.

CHAPTER ONE

Instrumental Faith

In the 1920s and 1930s, poor visibility was much more likely at the altitudes at which airplanes flew than it was at ground level due to the frequent occurrence of low clouds in the most populous regions of the United States. Early aircraft were restricted to altitudes of only a few thousand feet (at best), and thus they could not fly above the weather. They had no choice but to fly through it. For this reason, the blind flying problem was a more immediate economic concern than was the blind landing problem; fortunately, it proved a more tractable issue. By the mid-1930s, it had largely been solved.

There were two aspects to the blind flight problem: control of the aircraft in the absence of outside visual references, and navigation of the aircraft from one point to another without external vision. Innovation in point-to-point aerial navigation has been well documented elsewhere.[1] This chapter focuses primarily but not exclusively on the first aspect of blind flying, control of the aircraft. Before the late 1920s, the distinction had not yet been made. Many early pilots believed that their own motion senses provided all the information necessary to maintain control of the aircraft, and hence point-to-point navigation was considered the only problem. The real danger in blind flying, however, was not in getting lost. It was in inadvertently losing control of the aircraft. Before 1926, no one knew why this happened, and most fliers simply blamed it on insufficient skill among the unfortunate. There was a physiological reason for loss of control during blind flight, however, and it required a solution.

The solution was not primarily mechanical. The instrumentation necessary to resolve the blind flying problem already existed at the end of World War I, but for a decade after, no one used it for that purpose. Instead, pilots continued to fly as they always had—by the seats of their pants. In the late 1920s, a pair of U.S. Army officers, William Ocker and Carl Crane, began a campaign to convince their peers that this was no longer an acceptable method for professional pilots. Human motion senses did not function accurately in the three-dimensional world of flight, they claimed. Instead, these senses misled pilots and caused them, ultimately, to kill themselves. Based upon research they had done on other army pilots and on airmail service pilots, they contended that aviators had to ignore their motion senses and follow a set of rules built around their instruments in order to fly blind safely.

Their mechanistic version of flying ran directly counter to the prevailing mythology of flight that good pilots were born, not made. They met a great deal of resistance at first, but a series of events beginning with Charles Lindbergh's famous transatlantic flight gradually overcame professional pilots' resistance to the new rule-based method of blind flying. The most important such event was the airmail crisis of 1934, which caused the adoption of simulators to ensure that the new rules of flying were instilled as habit—without leaving the ground. Habituation was necessary because the central problem in blind flying was one of faith. Pilots had to learn to believe in their instruments rather than themselves. This was not an easy task.

INSTRUMENTS FOR BLIND FLYING

The Great War provided the impetus for the development of a wide range of aircraft instruments and aerial navigation techniques. Prewar aviation had been exclusively a fair-weather affair, and most flying was local. Aircraft took off from and landed on the same airfield, having gone no more than a few miles away. Pilots could use familiar landmarks for what little navigation they did. They could pick a landmark and fly toward it, then choose a new one for their next waypoint. They also made use of roads, railroads, and rivers as navigation aids by simply following them. This kind of flying became known as contact flying because the pilot remained in visual contact with the ground.

Contact flying proved to be unwise during the war for a number of reasons. Ground fire was one obvious problem. Even small arms could hit an aircraft flying close enough to the ground to see landmarks, and by the end of the war antiaircraft guns capable of reaching above ten thousand feet had been produced. The desire to stay alive drove pilots to fly too high to rely on landmark navigation. Even discounting the effect of gunfire, however, contact flying was not very useful. Contact flying worked best when a pilot was familiar with the route over which he was flying, but pilots in the Great War (especially American pilots in the war's final eighteen months) were expected to operate over hostile and unfamiliar terrain. There were no aeronautical charts of enemy territory, and charts of friendly territory were new and not yet widely available. Railroad maps became precious to pilots because the iron horses' tracks were permanent, visible from high altitudes even at night (their polished surfaces reflected moonlight), and always led somewhere. But following the railroads imposed limitations and dangers. They did not go everywhere along the front; hence relying on them limited potential targets. And if planes routinely followed the tracks, the enemy would quickly line the tracks with guns to shoot them down. The "iron compass" of the railroads provided a decent emergency navigation aid, but turning the airplane into a really useful weapon meant replacing contact flying with navigation less dependent on detailed knowledge of enemy terrain.

The obvious solution was the same kind of chart, compass, and stopwatch navigation that navies of the world had used since the fifteenth century—except that the now-ancient technology of the magnetic compass did not seem to work in the air.[2] More precisely, it worked in tethered balloons but not in airplanes, where it behaved erratically. Sometimes the compass seemed to "stick" and not respond at all to a turn, while at other times it responded too quickly and overshot the plane's actual heading. It might start to turn one direction when the plane was turning the other direction. The compass provided reliable indication only when the plane was in straight, level flight. This capricious behavior caused fliers to mistrust the magnetic compass and caused researchers to seek an explanation and a means of fixing the problem.

Keith Lucas, a researcher at the Royal Aircraft Factory, explained the compass's capriciousness in late 1914. The curvature of the earth's magnetic field was the culprit. Because the north-seeking end of the needle is pulled "down" toward the north pole (in the Northern Hemisphere), compass manufacturers placed the pivot point off center slightly to keep the needle from actually pointing down. As long as the compass was level (and high-quality compasses included bubble levels for precise reading), the needle stayed flat. But since airplanes turned by banking, thus raising one wing above the other, the compass was not level with respect to the earth's magnetic field during turns. As the plane rotated around the needle, the off-center pivot combined with the field's downward pull caused erratic movement. With Lucas's explanation at hand, the Royal Aircraft Factory was able to ameliorate, but not fix, the magnetic compass's unreliability.[3] Yet recognition that the magnetic compass could never be used as a turn indicator led investigators to try other techniques.

One approach was to detect the degree of an aircraft's bank. That proved simple to do, and by the end of the war a bank indicator consisting of a fluid level with a small metal sphere in it to mark the angle had been developed. While fine for detecting a bank, it proved of little use as an instrument to supplement the magnetic compass, partly because one can turn a plane without banking (properly called skidding), but mostly because it was not very precise. One could not use it to make small course changes because it did not detect small angles of bank reliably and because it was slow to respond in any case. Hence it was not a good turn indicator.

Because of its well-known ability to maintain a constant orientation in space, the gyroscope was the immediate favorite of inventors interested in making a turn indicator. The idea of a gyro turn indicator actually preceded the airplane, according to an internal memo from the Pioneer Instrument Company, a spin-off from the more famous Sperry Gyroscope Company. British patent #3587, dated February 17, 1899, covered the basic idea of a turn indicator and included drawings that functionally replicated the equipment that Pioneer Instrument began to produce in 1919. An air-jet powered gyroscope rotor was suspended in a frame that was free to rotate about an axis perpendicular to the axis of rotation. As the aircraft turned, it literally rolled around the gyro, and the geared frame moved an indicator needle that showed the divergence between the gyro's orientation and the aircraft's. Elmer Sperry's son Lawrence test flew the device in 1917 but found that the instrument was not responsive enough. Two years of work by Elmer and Charles H. Colvin produced one that Lawrence found acceptable. Elmer filed his patent for it in 1920, but at least a half-dozen other gyroscope-based turn indicators appeared by 1919. The Royal Aircraft Factory, for example, produced a detachable gyro turn indicator that was simply clamped somewhere convenient on the airframe, with the open gyro wheel powered directly by the air flowing past the aircraft, while the German Drexler "Steering Gauge" used an electrically powered gyro to do the same thing.[4]

Sperry Gyroscope spun off Pioneer Instrument Corporation in February 1919, when several of Sperry's employees decided they wanted to keep working in aeronautical instrumentation after Elmer Sperry proclaimed that there was no future in them. They were able to license the patent rights from Sperry to design the turn indicator shown in Figure 1.1. Pioneer received a substantial order for the turn indicator from the French l'Armée de l'Air in 1919, but the U.S. Army Air Service, a unit of the Army Signal Corps, did not follow the French lead.[5]

Yet the army's failure to adopt the instrument immediately did not matter because fliers had not yet recognized the turn indicator's most important use. It had been invented as a means to overcome the instability of the magnetic compass during a deliberate turn, which was how those pilots who had it used it. The turn instrument was a rate instrument, which meant that it displayed, in effect, how fast the aircraft was turning. One could therefore make turns accurately by placing the needle at a particular spot on the gauge face and then timing the turn. With practice, a pilot could learn to accurately make a turn of any size by this method. The use of the turn indicator was therefore based on intention—pilots used it when they intended to make a turn. Otherwise, they concentrated on the magnetic compass.

But fliers with the turn indicator kept dying, including some accomplished heroes. Capt. John Alcock, who made the first nonstop transatlantic flight in June 15, 1919, died in a fog-related crash that December, leading Elmer Sperry to write to Lawrence, "We have absolutely got to solve the problem; if we die in the attempt and could have registered a single notch in advance, it seems to me that it would be well worth while." Tragically, Lawrence did not live to see the problem solved, drowning in the English Channel on December 13, 1923, after a daylight crash.[6] But Elmer's attitude was hardly unusual. In the same letter, he quoted Charles Kettering: "We do not yet know this airplane business. One need only fly for a few months to realize that the airplane is developed to about the same point as if we had ships with absolutely no paraphernalia of navigation. . . . The only way we will bring [navigation paraphernalia] into existence is to try to do something; to find out what we can do and then do it. And that is why it is right to continue the airmails. That is the reason I believe in them."[7]

The airmails, as Kettering prophesied, were the laboratory in which blind flying was perfected. The Post Office did not work alone at this. Most of the hardware the Post Office tested was developed by the Army Air Service at Wright Field or by private companies. But the Post Office did run a small experimental facility at the Air Mail Service's Development Division at Monmouth Field, Illinois, where it orchestrated testing of these innovations.

At war's end, blind flying research work largely ceased in the U.S. Army until the mid-1920s, while the U.S. Post Office, which then prided itself on its technologically progressive nature, took up the mantle of aviation progress. The Post Office had begun canvassing Congress to get funding for an airmail experiment in 1910, without success. In early 1918, the Army Air Service had approached second assistant postmaster general Otto Praeger, who was lobbying hard to get airmail funding from the U.S. Congress, with an offer to fly the mail. The Air Service needed the experience with cross-country flying, and its officers believed that flying the mail was the way to get it. Praeger could hardly turn down the army's

Figure 1.1. Pioneer Instruments turn indicator. National Air and Space Museum, Smithsonian Institution (SI 83-7649).

offer, given the patriotic fervor of the day, and the army began flying the mail in May. The army's leadership quickly abandoned the project however, and by August had turned over pilots and aircraft to the Post Office.[8]

Praeger believed that the army had failed because it was not devoted enough to the airmail cause. Although not a pilot, he was a true believer in the airplane's great potential. Like Kettering, he believed that only the government could develop the airplane to a point at which it could be commercially viable. He expected that the airmail would lose money and pilots for years, but like the Sperrys, he believed the sacrifice was necessary to bring about the aerial age.

Praeger's belief in the airplane's potential, coupled with his own lack of personal knowledge of the technology's current limitations, led him to place enormous pressure on his pilots to fly in all conditions. Praeger had suspected that the army's pilots routinely had canceled flights for bad weather when the trip was actually flyable. He forced the mail pilots to sign contracts that specified their responsibility to fly in fair weather or foul, and acted on them. When four pilots refused to fly the New York to Washington, DC, run on July 22, 1919, because visibility was less than two hundred feet, he ordered the two senior pilots fired and the two junior pilots reprimanded.[9] This action came at the end of a month in which two aircraft had been demolished in weather-related accidents from which the pilots had walked away and three days after the death of a pilot who crashed after becoming disoriented in cloud cover. The New York pilots initiated a brief strike over the firings which was resolved when Praeger reinstated one of the two pilots and granted airfield managers the authority to determine flyability (thereby removing the authority from the pilots and from himself). The tone was set, however, and the pilots' contracts continued to state that refusal to fly constituted their resignation.

Praeger's successors were somewhat more safety minded, but the pressure to fly in all weather was endemic to the enterprise. Businesses depended on the mail more for reliability than speed, and unless the airplane could be made to perform on schedule, it could never compete with the older, better established, and more reliable forms of mail transportation. Its sole advantage over those forms of transportation was its superior speed, but that counted for nothing if the plane had to stop at night or for bad weather. Express trains, the primary means of carrying mail in the first half of the century, were slower but stopped for almost nothing. During the long winter nights especially, the trains could overtake and pass the faster, but grounded, mail planes. Hence the mail planes did not start to beat the express trains routinely on the New York to San Francisco route until they started to fly the segment from Chicago to Cheyenne, Wyoming, at night in 1923.[10]

The airmail service did not become profitable for the Post Office until it began flying the New York to Chicago route at night in 1925. A group of Chicago bankers had carefully explained to Praeger's successor, Paul Henderson, that because the airmail left both cities in the morning and arrived in the evening, it still arrived in the banks the next morning along with the mail that went by train. There was no speed advantage from the banks' point of view and thus no point in paying the higher airmail postage. If the mail left at night, however, and arrived the next day, it would essentially eliminate the transit time for bank drafts (since the banks were closed at night), saving the banks millions of dollars in "float" costs per year.[11] This savings would make airmail well worth the cost. The Post Office listened, and its airmail volume between New York and Chicago exploded during the last two years it flew the mail before privatization.

Hence the airplane's miles-per-hour speed was irrelevant if it did not translate into faster service than that provided by the railroads. That, in turn, meant all-weather, round-the-clock operation. Kettering, Sperry, Praeger, Henderson, and a few other entrepreneurs intended to use the airmail to develop the technologies and techniques to permit all-weather flying. In essence, the airmail was a real-world laboratory for flying. Most significant about the airmail experience was not new technologies; in fact, the Post Office developed and deployed no instruments or other navigation technologies that had not existed during World War I. What it did was build an infrastructure to support flying, while its pilots began to build a new kind of flying around the instruments wartime development had bequeathed them.

That a new kind of flying was necessary was not immediately obvious to the Post Office's pilots, and most of them continued to fly the way they always had. Quite a number of them died doing it, and more crashed but lived. There were essentially two kinds of weather-related accidents during the airmail years. The first type, flight into terrain, were caused when pilots in poor weather descended to very low altitudes to keep the ground and its handy navigation aids in sight. On October 14, 1919, for example, pilot Lyman Doty was flying the Washington to New York route when he descended below treetop level in an attempt to keep the ground in sight. He hit a tree, the plane caught fire, and he burned to death. On March 10, 1920, Clayton Stoner was the victim of another in-flight collision with a tree on the Chicago to Cleveland route, and later that month Harry Sherlock hit a smokestack while approaching Newark's Heller Field. While flying in four-hundred-foot visibility from Chicago to Cleveland in April 1921, J. Titus Christensen was following the Cuyahoga River when he entered a cell of clear air over the river, which was surrounded by tall buildings and bridges on all sides. He tried to climb out in a tight circle, staying inside the clear cell, but stalled the plane and crashed into railroad tracks.[12] Trying to contact fly in bad visibility was extremely hazardous.

Yet the obvious alternative, staying up high in the clouds where trees and buildings did not reach, was perhaps even more frightening. Many of the mail pilots who tried this alternative died too. While trying to fly over the Snowshoe range of the Allegheny Mountains in Ohio in 1919, Charles Lamborn's plane entered the clouds only to reappear a few minutes later, nose down, and plummet into the ground. John Charlton died the same way that year, apparently after spinning into a mountaintop. On January 4, 1923, Henry Boonstra tried to fly over a storm at eighteen thousand feet when his engine died. Trying to glide down through the clouds, he lost control of the airplane and went into spins five times before finally hitting the ground.[13] Fortunately, he lived to tell the story. A lot of other pilots experienced these spins in clouds and recovered from them when the spin dropped them to lower altitudes with a visible horizon.

The mail pilots found these spins to be deeply frustrating because they could not explain the phenomenon. Their aircraft were aerodynamically stable, meaning that the airplane would return without pilot action to straight, level flight after a small disturbance. Thus the pilots were becoming disoriented and causing the spins themselves, but they did not know why. Most early mail pilots chose to take the collision risk of flying low rather than to face the spin risk of flying blind. The pilots who chose to face the spin threat, however, were the ones who ultimately produced the new kind of flying.

The disorientation problem, they learned from anatomists, was a result of how humans sensed motion. Beginning in 1926, two army officers, pilot Capt. William C. Ocker and flight surgeon Maj. David Meyers began a campaign to convince pilots that the error lay in believing that people could fly using their sense of balance. People maintained their equilibrium through a combination of the fluid levels in the canals of the inner ear, through muscle balance, and through vision. Deprived of vision by clouds, the other two senses became misleading, causing pilots to make inappropriate control actions. The two men had employed a Jones-Barany chair, essentially a barber's chair, to demonstrate that "after accelerations, either angular or linear, a man cannot competently interpret the movements of his body unless he can see some plane of reference such as the earth or objects stationary in relation to it."[14] By spinning pilots in the chair a few moments at a constant speed, then slowing it, Ocker and Meyers demonstrated that their subjects believed that they were now spinning in the opposite direction.

This effect, to which they applied the old term for dizziness, *vertigo*, was what caused pilots to kill themselves. If a plane started into a turn, a pilot could sense it accurately. If he then applied the proper control corrections to stop the nascent turn, he would sense that the plane was now turning in the opposite direction when it was not actually doing so. Correcting for that apparent motion led to further inaccurate sensations of motion and further inappropriate control movements. After only a few minutes, the difference between what a pilot thought the plane was doing and what it was actually doing became so great that the pilot lost control. In reality, as the gap between apparent and actual motion grew, pilots began to overcontrol, eventually making too tight a turn and stalling the plane, which in turn caused a spin.

Between 1927 and 1932, when Ocker and army pilot Carl Crane published their full study in a highly influential book, they tested a large number of mail, airline, and army pilots to find that only 3 percent could fly blind for more than twenty minutes without entering a spin.[15] Even when equipped with the turn indicator, pilots would revert to believing their sense of balance in flight and mistrust the instruments that claimed something counter to their own senses. Ocker, Meyers, and Crane therefore promoted a method of blind flying that was based on complete dependence upon the instruments.

The Ocker-Meyers method of blind flight was based on the idea that pilots should create in their minds a virtual artificial horizon, based on the information presented by their instruments.[16] As no mechanical artificial horizon existed in the United States before 1929, they believed that a virtual horizon was necessary to produce safe blind flight. Constructing such a horizon mentally based upon the turn-and-bank indicator, the climb indicator, and the compass took great concentration. The magnitude of concentration required to create and maintain that mental image of the horizon induced fatigue fairly quickly in most pilots, and only a very few (like Charles Lindbergh, an early convert to the method) could maintain it for hours at a time. This difficulty helped generate a great deal of controversy over their method in the aeronautical journals, and Ocker and his supporters used their Jones-Barany chair to prove their case.

Ocker built and patented what was literally a black box positioned over the subject's eyes while seated in the Jones-Barany chair.[17] The box contained the standard combined turn-and-bank indicator with an air pump to keep the gyro spinning and a magnetic compass. For the first test run with a subject, the box's lights were turned off so that the subject was literally blind, and the subject was spun in the chair to demonstrate the vertigo effect that prevented him from accurately sensing motion. The second test run was done with the box's lights on and the turn indicator's gyro spun up, to demonstrate that the turn indicator often opposed the subject's own sense of motion. The turn indicator was always correct— the demonstration depended upon pilots recognizing that a gyroscope could not be wrong.

Acceptance of the Ocker-Meyers method of blind flight did not come quickly. Their earliest articles in 1926 produced controversy in the flying community because most (nonmail) pilots believed completely in their ability to fly blind without the luxury of turn-and-bank indicators or the offensive idea that pilots should focus completely on instruments. Most pilots flew for the joy of seeing the earth from the air and had no desire to place their attention inside the cockpit. Many of these amateur pilots had encountered blind conditions briefly and survived them, after all, a reality that Ocker admitted readily. The large majority of pilots placed themselves in a class that Ocker and his advocates called "natural pilots," who flew by sense. They did not wish to become, and did not believe it necessary to become, "mechanical pilots" who flew by memorized rules built around instruments.[18]

Even pilots who had accepted the Ocker-Meyers method did not particularly care for it. The sheer difficulty in constructing and maintaining a virtual artificial horizon caused professional pilots to return to the idea of developing a real instrument to do this for them. Ocker and Crane promoted a "flight integrator" that was supposed to solve this problem, but they never described it in much detail. In any case, a mechanical artificial horizon was not a new idea in the late 1920s. In 1916, the U.S. Navy's Office of Naval Aeronautics had asked Sperry for a gyroscopic compass that established directional and horizontal planes of reference. Elmer Sperry had responded that they needed two instruments to accomplish this, a gyrocompass and a gyroscopic artificial horizon.[19] Sperry failed to produce either device during the war, and the navy dropped the contract at war's end.

The appearance of a German artificial horizon instrument called the Gyrorector in 1926 helped stimulate new U.S. interest. The army obtained one and tested it at Wright Field for several days before declaring its inadequacy for army use.[20] Its chief problems, according to the army's report, were that it was too slow to respond and much too heavy (over thirty pounds). Weight in 1920s aircraft was a major issue. Adding a thirty-pound instrument to the mail planes of the day reduced the amount of carriable mail by 10 percent, and the mail planes were simply modified army planes. The army wanted something a good deal lighter. The slow response of the gyrorector was also a substantial operational drawback, as pilots ideally wanted an instrument that responded instantly. Ultimately, this was the easier problem to fix. All instruments have to be damped in some fashion to prevent unwanted oscillations, a reality known since the nineteenth century. Too much damping and the instrument responds slowly; too little and it will appear wild to the user. Striking the proper balance between speed of response and stability is less a technical problem than a cultural one. The instrument designer has to understand where the user's preference lies and build that into the instrument. That was a simpler issue than shaving twenty pounds off the device without making it too fragile to survive flying.

Fragility had doomed Elmer Sperry's earlier work on the artificial horizon. In his wartime foray into aeronautical instruments, he had discovered that the acceleration forces endemic to flight were much greater than those encountered by the seagoing gyroscopes in which he had previously specialized.[21] Worse, because airplanes were extremely weight-limited, the instruments had to be much lighter. That closed off the obvious solution of making the instruments more rugged by making them more massive. They had to be less massive and more rugged, a combination that Sperry did not master during the war.

Sperry received a chance to redeem his failure in 1927. Early in the year, Daniel

Guggenheim established a fund, run by his son Harry, devoted to the improvement of aeronautics. This Guggenheim Fund for the Promotion of Aeronautics
financed schools of aeronautics at the University of Michigan, the Massachusetts
Institute of Technology (MIT), the California Institute of Technology (CalTech),
Stanford, the University of Washington, and the Georgia School of Technology,
while undertaking a real-world project to perfect blind flying. The fund arranged
to borrow aircraft from the army, hired army reserve pilot James "Jimmy" Doolittle, and set up an experimental "Full Flight Laboratory" at Mitchell Field, New
York, to carry out its blind flying research using radio equipment borrowed from
the National Bureau of Standards, Bell Labs, and Western Electric. The goal was
to produce a complete blind flight, including a blind landing. Doolittle ran the
blind flying effort at Mitchell Field, assisted by MIT professor William G. Brown.
Doolittle's recognition of the limitations of the turn-and-bank indicator and his
belief in the rightness of the Ocker-Meyers method led him back to Elmer Sperry.
Sperry decided what Doolittle needed was a horizon indicator and turned the task
over to his son Elmer Jr.[22]

The instrument Elmer Jr. devised used an air-spun gyro designed to maintain
a vertical orientation. The instrument face was designed to replicate the natural
horizon, with a fixed set of wings representing the aircraft and a "horizon" line
that appeared to move in relation to the wings. The upper half of the instrument
face was light blue, to indicate skyward, and the lower half was black to represent
the earth (Figure 1.2). For the Guggenheim blind flying tests, the new device was
mounted in a Consolidated NY2 borrowed from the army.[23]

Doolittle made his first complete blind flight on September 29, 1929, taking
off from and landing at Mitchell Field. Using the horizon, a directional gyroscope,
a specially designed altimeter, and a set of borrowed radio equipment, Doolittle
took off under the hood like that in Figure 1.3 and flew a well-practiced track away
from and then back toward the field. He landed using an experimental blind landing system devised by the National Bureau of Standards (discussed in Chapter
3).[24] For his effort toward making blind flying possible, Doolittle received the
1929 Collier Trophy.

The Sperry artificial horizon was rapidly adopted by the airlines, whose pilots
found that it made keeping the plane straight and level much easier. This was because the artificial horizon sensed the absolute amount of deviation of the wings
from true horizontal, while the turn indicator displayed the plane's rate of turn.
A slight bank showed up on the artificial horizon instantly, while the same bank
might not translate into a turn rate substantial enough to show on the turn indicator for several minutes. Although many promoters of the artificial horizon ad-

Figure 1.2. Sperry artificial horizon. Reprinted with the permission of Unisys Corporation, courtesy National Air and Space Museum, Smithsonian Institution (SI 99-41447).

Figure 1.3. Boeing School of Aeronautics blind flight trainer. Note the leather hood over the front cockpit. Blind flight meant *blind* flight. National Air and Space Museum, Smithsonian Institution (SI 99-41453).

vocated it as a replacement for the turn-and-bank indicator, most airlines put both on their aircraft once the horizon became available. Pilots found that while the horizon made maintaining level flight easier, the turn-and-bank indicator promoted simpler recovery when the aircraft departed from level flight.

Simply telling pilots how to use the instruments and sending them up did not solve the problem. Pilots tended to believe that the turn indicator and artificial horizon worked fine as long as they could see, but once in the clouds, they thought the instruments went haywire and reported turns that the pilots were certain the airplane was not making.[25] The turns did happen, of course; that was the point. Pilots had to be broken of their dependence on their sense of balance.

The first step toward freeing pilots from this dangerous dependence upon themselves came from the army, which began screening its new volunteers for flying duty for instrument compatibility. Once a potential recruit had been taught the basics of flying, he was given two hours of classroom instruction on how the instruments worked and why they were necessary. An instructor then took him aloft in a training aircraft with a hood fitted over the student's cockpit, and once high enough for safety turned the plane over to the student. If the student tried to maintain control via the instruments, he was acceptable. If he did not, he was given two more chances after additional classroom instruction. If on the third training hop he continued to rely on his own sense of balance, he was washed out of the flight program. Henceforth, willingness to rely completely on instrumentation was a fundamental requirement for army fliers.

The airlines went further in checking out all of their existing pilots, motivated in part by a Department of Commerce ruling that all airline pilots had to pass a blind flying test by January 1933.[26] Eastern Air Transport, for example, sent Howard Stark around their system to retrain all of the company's pilots. By 1930, Stark was a recognized expert in blind flying because he had improved upon the Ocker and Meyers method. While they had advocated that pilots learn to fly by instruments alone, Stark had devised an explicit methodology for using the instruments in blind flying. Called the 1-2-3 method by Stark (and the A-B-C method by the army), it was rapidly adopted in both army and airline flying.

Stark had begun his flying career as a barnstormer and joined one of the new airlines in 1927. He quickly learned that he did not really know how to fly; when he asked his fellow pilots how to use the instruments he was told to "keep them centered." Keeping them centered proved not to be easy. After entering a spin at four thousand feet and plunging to less than seven hundred feet before recovering, he realized that "he really could not keep control of the plane if [he lost his] balance, and this might happen to anyone in rough air such as thunderstorms or line squalls."[27]

After his personal epiphany, Stark developed his 1-2-3 method during 1928 and 1929. The steps of the method were: (1) center the turn indicator with rudder only; (2) center bank indicator with ailerons only; and (3) center climb indicator with elevators only. The order in which pilots were to take these steps mattered, too, he explained in his 1931 textbook. If the turn indicator was not centered first, the bank indicator would "give a false index for level flight."[28] The method relied upon the realization that in a properly banked turn, which by definition is a turn intended to exactly mimic the acceleration of gravity, the bank indicator would indicate no bank at all and the pilot would not feel the turn through the inner ear's balance function, because neither of these sensors can discriminate between gravitational acceleration and centrifugal acceleration. By using the rudder to center the turn indicator first, a pilot would automatically create an improperly banked turn that would permit the bank indicator to deflect. Only then could the pilot level the aircraft's wings using the bank indicator as a reference. In essence, the bank indicator only accurately represented the position of the wings when the plane was *not* turning. This is why the order of the instruments and the steps related to them mattered. The turn had to be stopped before the bank indicator could be relied upon.

The turn-and-bank indicator had assumed a new function in Stark's method. While it had been designed to permit pilots to make accurate turns, supplementing the unreliable magnetic compass, in this method it became an instrument used to prevent the plane from making unwanted turns. Originally, the turn indicator was used only occasionally by pilots when they wanted to turn, but now pilots had to pay near-constant attention to it. Stark was not the first to perceive this use; Lawrence Sperry seems to have understood this use of the instrument, as did the old hands from whom Stark had asked advice when he took the mail job. Realization that the turn indicator could be used this way was probably widespread among professional pilots by the mid-1920s, but their success at using it varied greatly. It seems likely that many individuals had worked out their own way to use the turn-and-bank indicator, and Stark may well have codified something that was already common knowledge among professional fliers. By a year after his book's publication, however, his method was explicitly adopted at flying schools, by airlines, and by the army.[29]

Stark's method also separated the aircraft's controls in pilots' minds. That separation was a false one, and a later, more complicated method was proposed to restore a more realistic perception of control actions. By telling pilots to use the rudder to manage the turn indicator and the ailerons to manage the bank indicator, Stark was implicitly arguing that the rudder was for turning and the ailerons for banking, but actually the rudder causes banking in addition to turning, while the ailerons cause turning in addition to banking. In fact, only raw beginners use the rudder to make a turn; the ailerons are far more effective. By the late 1930s, designers were making airplanes without rudders, although the trend did not catch on among pilots, in part because of their training and in part because the rudder was important in emergencies like spins and loss of an engine in multiengine aircraft. Stark's rules removed the confusion that two controls with similar effects could create in pilots' minds. They were, in effect, a heuristic approach designed to produce consistent results, even if the process of applying it was inelegant and suboptimal.

Because of the need for pilots to continuously refer to the turn indicator to prevent turns, much of Stark's short book was devoted to the subject of proper instrument placement. For most effective use of his method, he believed, the "turn indicator group" of instruments should be placed in the following order: airspeed indicator, turn indicator (which included the bank indicator), climb indicator, and altimeter. Pilots could scan this set of instruments left to right, centering each one in its proper turn.

The introduction of the Sperry artificial horizon two years later led him to revise his recommendations only slightly. While pilots were to use the turn indicator group to recover straight and level flight after a disturbance caused the plane to turn, a new "Sperry group" that included the horizon and the directional gyro or gyrocompass was to be used to maintain straight and level flight. The Sperry group was to be in a specific layout, too. If only a Sperry group was to be installed, Stark advocated placing the horizon atop an inverted T, with the gyrocompass directly below it, and airspeed indicator and altimeter to the gyrocompass's left and right, respectively. The horizon's position atop the layout (where it would be directly under the magnetic compass, which was always placed above the instrument panel to limit electromagnetic interference), reflected its new status as the most used instrument.

Just as Ocker and Meyers had found advocates willing to promote their view of the incapacity of humans to fly blind unassisted by instruments, Stark found willing champions of his rule-based method. Ernest A. Cutrell, who had flown for the Department of Commerce in the early 1930s and was hired away to be an instrument instructor for American Airlines, advocated Stark's method to *Aviation*'s readership. At New York University, E. B. Schaefer from the Guggenheim School of Aeronautics wrote a detailed exegesis of it for *Aviation Engineering*.

Ocker himself gave the biggest impetus to Stark's method by republishing it (under a new name) in his co-authored 1932 book, which became the most influential text on instrument flying before World War II. By 1938, a former mail pilot and instrument instructor for American Airlines could claim that "instrument flying is based on cold logic; there is nothing of the 'born pilot' about it" before explaining how to use Stark's 1-2-3 method.[30] The mechanistic method of flying that Stark had derived during his airmail experience had come to dominate commercial and military aviation.

Stark's new kind of flying was difficult enough that the airlines had to set up training programs to teach it. Pan American, for example, set up a school in Brownsville, Texas. The school employed the same equipment that Ocker and Meyers had used to demonstrate the disorientation effect on the ground, the Jones-Barany chair and Ocker's black box, to familiarize its pilots with the problem. Pan Am also equipped an airplane (a Fairchild FC-2) with a second, blacked-out cockpit behind the original pilot's position and used that for initial in-flight training. Once student pilots could fly blind in this smaller aircraft, they were advanced to training in larger multiengine aircraft. Pan Am required a minimum of ten hours of instruction. It also required what would eventually be called *recurrent training*, making pilots redemonstrate their abilities at regular intervals.[31] Like United Air Lines, which instituted a similar program, Pan Am required an hour of practice monthly after students graduated from its course. The monthly practice was necessary to keep pilots who did not routinely encounter blind conditions from backsliding after the training course and returning to dependence on their sense of balance.

Stark's rule-based flying led directly to a radical idea, perhaps best stated by American Airlines instructor Karl Day in a 1938 textbook, that pilots should be taught to fly on instruments *first*, preferably while sitting on the ground.[32] Contact flying taught habits that might be acceptable for amateur pilots, but professional pilots needed to learn to reject the misleading siren calls of their motion senses and accept the truth as revealed by the instruments. He believed that instead of permitting pilots to gain hundreds of hours of experience in contact flying before being trained on instruments, they should be taught instrument flight before learning contact flying or, second best, immediately after they successfully completed basic flight training. Day took the teachings of Ocker, Meyers, Crane, and Stark to their logical conclusion. If instrument flying was to be based on rules, those rules should be ingrained in pilots as habits, and training programs should be arranged so that proper habits were formed early.

Training nonpilots to fly on instruments first, however, was an idea whose time never came, in part because of the danger to instructors imposed by a completely untrained student at controls enclosed in a small, dark space in which he might panic. The idea of training new graduates from flight school on instrument flying was adopted by the Army Air Corps, which appended instrument flight training to its basic flight school curriculum. It did not catch on among private fliers, however, because instrument training was very expensive compared to basic flight training. The ten hours of instrument training that the airlines and the army considered necessary to produce a proficient instrument pilot was more than the total amount of flying time required for many students to receive their private pilots' license during the 1930s. Physicist Luis Alvarez, for example, received three hours of flight training before soloing.[33] The Bureau of Air Commerce, which was founded in 1928 to regulate civil aviation, was not about to require a sudden doubling or tripling of the training requirements for all pilots. The organization rightly perceived that doing so would be political suicide. Only commercial and military pilots saw their training requirements increase during the 1930s to reflect the increased importance of "mechanical flying."

The cost of that increased training requirement led indirectly to the fulfillment of Day's claim that instrument training was best done on the ground. Day was not the first to believe so, and by the time he wrote his textbook, ground-based training was old news, if not yet famous. The idea of ground-based flight training was around even before the Great War, embodied in, for example, the 1910 "Sanders Teacher."[34] But ground-based flight training did not catch on until the ideology of instrument flight was accepted in the army and the airlines. It made little sense to adopt ground training for pilots when most pilots flew for the joy of the act of flying. Most people in the 1920s learned to fly from itinerant barnstormers after a handful of hours of instruction, and therefore organized flight schools were rare before the early 1930s, when the Bureau of Air Commerce began to formalize, if not tighten, licensing requirements. There was thus no market for ground trainers, regardless of their functionality. There were, nonetheless, a number of such systems invented in the early years of aviation, but only one happened to coincide with the rise of instrument flying.

The lucky inventor was Edwin A. Link Jr., the son of a manufacturer of player pianos. Born in Huntington, Indiana, in 1904, Link fell in love with the airplane during his youth. He took his first flight in 1920 with Charlie Chaplin's brother Sidney, and abandoned prep school in favor of a more hands-on education, which he got at his father's factory, then in Binghamton, New York. He devoted his spare time to tinkering with electronics and learning to fly, and he finally soloed in 1926. He purchased one of the new Cessna Aircraft Company's first planes the

Figure 1.4. Ed Link posing with his "blue box," an early blind flight trainer. National Air and Space Museum, Smithsonian Institution (SI 99-4151).

next year. But like many previous inventors, Link thought that flight training was too expensive. If the first few hours of "flight time" could be done on the ground, he thought training could be made less expensive and safer.[35]

He was encouraged in this belief when he discovered that the French had trained their pilots during the Great War by teaching them how to control the plane while taxiing around an airfield. This "penguin" system used planes with worn-out engines and clipped wings that would not allow takeoff. It also seemed to reflect his own experience, and he began to construct a trainer that would feel like an aircraft bouncing around an airfield. A year and a half of Link's spare time went into constructing his device which, when completed in 1929, included no blind flying instruments (Figure 1.4). It was merely a device intended to help prospective pilots get the feel of an aircraft's controls.[36]

As Link's biographer reports, almost no one in aviation wanted his trainer when it was introduced. Pioneer Instruments bought one to demonstrate its instruments to potential customers on the ground, and the U.S. Navy bought one to try out at its flight training school at Pensacola, Florida. Amusement parks and

fairs showed far more enthusiasm. Link sold fifty trainers to the amusements industry before 1933.[37] It took a disaster to introduce the professional flying community to Link's device, and that disaster descended from the airmail.

In February 1925, the U.S. Congress had passed the Air Mail Act, which directed the Post Office to let contracts to private companies to carry the airmail.
There were no airlines at the time; instead, during the year groups of investors
proposed airlines on paper in order to participate in the bidding. The postmaster
general awarded contracts that October to five new airlines—Colonial Air Transport, Robertson Aircraft Corporation, National Air Transport, Western Air Express, and Varney Air Lines—to serve routes that the Post Office's mail service
did not cover. It continued to fly the major trunk route (New York to San Francisco via Chicago) with its own aircraft. Once the new airlines had proven their
ability to maintain scheduled service, the Post Office opened bidding for this final
route in 1927, which it divided into two contracts. One contract covered the New
York to Chicago segment, which was awarded to National Air Transport; the other,
which went to Boeing Air Transport, covered the long Chicago–San Francisco
run. The Post Office also added new contract routes that year, mostly to new startups.[38]

The result of the bidding process, however, was a crazy quilt of routes. For a
piece of mail to get from Boston to Seattle in 1928, for example, it flew from
Boston to New York on Colonial Air Transport, then had to be put on a National
Air Transport plane to fly to Chicago. Once in Chicago, the letter was transferred
to a Boeing Air Transport Plane, was flown to San Francisco, and then went on a
Pacific Air Transport plane up to Seattle. This made no business sense at all. President Hoover's postmaster general, Walter Folger Brown, decided to rationalize
the mail service through a set of "shotgun marriages." Out of Brown's weddings
came the airlines that dominated American commercial aviation through deregulation in the late 1970s: American Air Line, Eastern Air Transport, Northwest
Air Lines, Transcontinental and Western Air (TWA), and United Air Line. Much
of this manipulation occurred at a single conference that began May 19, 1930, in
Washington, known ever after as the "Spoils Conference."[39] The conference produced a relatively rational, efficient route system based on a few large carriers,
while effectively locking smaller airlines, especially new startups, out of the lu-
crative mail business.

Franklin Delano Roosevelt held rather different ideas about what constituted
competition, however, and after his election he chose to make an example of what
critics had begun to call the "airline trusts." Roosevelt was driven to act against
the airlines in February 1934 by an investigation started by a political opponent,

Senator Hugo Black, who had received a distorted version of Brown's work from a reporter. The reporter had been approached by several airline hopefuls foiled by Brown's focus on constructing a few large carriers. He had prepared an exposé, only to find it suppressed by his editors. Black's investigation made the spoils conference public, prompting Roosevelt to act precipitously. On February 9, Roosevelt asked Maj. Gen. Benjamin Foulois, chief of the Army Air Corps, if his organization was prepared to fly the mail. Foulois had answered that it was, and Roosevelt directed Postmaster General James Farley to cancel all the airmail contracts.[40]

The result was a disaster but proved to be Ed Link's opportunity. The airlines (except United) received the majority of their revenue from the mail contracts, and with the exception of United and American, they shut down immediately in hopes of fending off bankruptcy until the political situation could be repaired. All the airlines launched an immediate media offensive against Roosevelt, and they were aided immeasurably, although certainly not deliberately, by the Army Air Corps.

The Air Corps failed Roosevelt catastrophically. The Chief of the Air Corps had assured Roosevelt that they "had a great deal of experience in flying at night, and in flying in fogs and bad weather, in blind flying, and in flying under all other conditions." Yet over the thirty days following the February 19 cancellation, ten army fliers died in the Air Corps attempt to imitate the airlines. The newspapers turned their deaths into a vicious political attack on Roosevelt, and the incident became a major political crisis. On March 10, the Air Corps called a safety halt and resumed flying the mail, only in fair weather and by daylight, on the nineteenth. The airlines got the mail back in mid-May, after being forced to change their names slightly (United Air Line became United Air Lines) in order to bid for new contracts. The rebidding did result in contracts going to a few new competitors, primarily Braniff and Delta, so the cancellation had not been an entirely futile act.[41]

The Air Corps failure at the airmail came not from a sudden increase in deaths; in fact, its attempt to fly the mail produced the same accident rate its normal operations in previous years had. But that rate was far worse than the airlines', and many of the accidents came specifically from failing at blind flying. The Air Corps used the same instruments the airlines did and had the same procedures described by Ocker and Crane's text. What its pilots lacked was proficiency. In its normal operations, the corps had no reason to fly blind. As numerous historians of air power have noted, Air Corps war-fighting doctrine was predicated on desertlike weather, clear and dry. One could not bomb a target effectively if one could not see it, and a different version of the same incontrovertible rule held for fighters (then called *pursuits*). Air combat required being able to see the enemy at a distance of miles until the development of radar, still years away in the United States. The corps had no reason to practice blind flying routinely, and it did not. Flight time was expensive, after all, and the corps was the poorest of the military branches by far.

Although Foulois had assured Roosevelt of the corps's competence, either he or some of his subordinates had been considerably less certain of the Air Corps's capabilities. One of Link's marketing partners was invited to demonstrate Link's trainer at the Casey Jones flying school in Newark on the eleventh, a scant two days after the cancellation announcement but several days before the corps started mail service. The Air Corps officers left impressed with the trainer's potential to provide inexpensive proficiency training, but they had no money to buy it. Link's partners canvassed Congress to rouse political support for an emergency appropriation to enable an Air Corps purchase of ten and, aided by the Corps' failure in the following days, succeeded. Roosevelt signed the appropriation bill in late March, and the first six trainers were delivered in June.[42] The Air Corps published new training procedures built around the Link trainer the following year.

The Air Corps's purchase served as validation of Link's trainer for other aviation groups. Japan was the first to buy (another ten), followed by the Soviet Union's ubiquitous Amtorg Trading Company. Orders from the United Kingdom, France, and various airlines followed, and Link was soon able to begin reselling to the same customers by introducing new models, often tailored to represent specific aircraft. Thirty-five countries used Link's trainers to train their pilots and to maintain their instrument proficiency by the eve of World War II.

CONCLUSION

By 1940, professional pilots had accepted the once-heretical notion that blind flying required complete faith in their instruments. Their organizations had institutionalized training programs to induce and maintain that faith, based around a machine whose own foundational concept—that pilots should be taught to fly on the ground—was not popular in the community's early years. Finally, professional pilots had adapted to a new kind of flying that replaced intuition and feel with instruments and rigid rules about how to use them.

By the late 1930s, therefore, blind flight was possible for pilots possessing the proper training and skills, although it remained dangerous for those without them—or for those who were simply out of practice. Success, however, had required instilling in pilots an extreme form of technological faith that required them to reject the sensory motion data upon which they had previously relied. Maintaining that faith meant ongoing retraining in mechanical flying.

Because blind flying depended upon pilots' faith in their instruments, achievement of blind landing necessitated development of a highly reliable system. A blind landing system would have to work in all weather (to support training in good weather as well as bad). It also would have to earn pilots' trust. And it would have to function within the existing aviation infrastructure.

CHAPTER TWO

Places to Land Blind

While the goal of all-weather operations forced pilots to learn to fly blind, it also caused promoters of air commerce to seek landing fields capable of supporting all-weather flying. Airfields during the Great War had simply been grass fields, usually circular, and aircraft took off and landed in whichever direction happened to be into the wind. But during the 1920s, airfield operators wishing to attract the mail business began laying surfaced runways, usually in the two or three prevailing wind directions at their individual sites. Why they did this is a matter of some dispute in aviation history, and we must examine the shift to runways in the context of the problems they created for blind landing technologies.

The transition from grass field to paved runways is usually linked anecdotally to the introduction in the early 1930s of heavy, all-metal airliners like the Boeing 247 and the Douglas DC-3. The chief difficulty with this argument is that airports began adopting surfaced runways well before those aircraft were developed. A more recent interpretation places the responsibility for the adoption of surfaced runways on architects, whose professional desires drove them to promote the permanent construction of geographic forms in the ground.[1] Unfortunately, this explanation reflects the popular sin of social determinism—architects' professional desires no doubt affected runway layouts, but they were not the reason airfield managers and airlines wanted grass replaced in the first place. This argument cannot explain why airport managers wrote many letters to journals of the day explicating detailed physical problems at their airports. Runways emerged as a solution to specific problems that various airports had in meeting demands for regular service. Surfaced runways were invented as a solution to particular local environmental problems that prevented the achievement of regularity.[2] They are thus part of the same drive toward "all-weather operations" that encompassed blind flying and landing efforts.

The deployment of surfaced runways, however, led to a substantial increase in the difficulty of designing an acceptable blind landing system. Making a blind landing system capable of guiding a plane into a large, open grass field was one thing, while designing one to guide it onto a narrow strip of pavement proved to be quite another. Runways required greater precision than did the grass fields of the Great War. Hence, an early blind landing system developed to serve grass fields proved unacceptable to pilots and airport operators.

WHY RUNWAYS?

The earliest airfields in the United States were unprepared in any way save clearing of trees and shrubs. The U.S. Army used parade grounds as takeoff and landing areas with its first aircraft but quickly found that marching soldiers and landing airplanes were not a good mix. The First Aero Squadron's initial operational deployment, during the U.S. campaign to track down Francisco "Pancho" Villa, cast the problem of landing fields into stark relief. As Benjamin Foulois reported in his autobiography, the soldiers sent to find landing fields for his squadron often chose poorly. Although naturally flat fields of sufficient size abounded in the campaign area, they were not smooth enough for safe operations. Rocks, holes, and even large clumps of grass caused crackups.[3] Although few serious injuries resulted, the damage to aircraft was expensive, in terms of repair costs and in reduced availability of the aircraft. The unprepared field was therefore obsolescent by the time of U.S. entry into World War I, used only for emergency landings. Proper airfields had to be leveled and graded.

The Army Air Service's *Specifications for Municipal Landing Fields,* published in 1919, was the earliest basis for airfield design in the United States.[4] The army identified the square as the best shape for an airfield based on its experience during the Great War.[5] A first class field, in the army's opinion, was a square 1,800 feet by 1,800 feet, with a central 150-foot concrete "runway" in the shape of a cross. The runway was for takeoff, not landing; indeed few army fields were built with this feature. Instead, most army airfields remained what were called "all-over" fields (with no specific runways) through the mid-1930s. They were groomed and drained turf fields that allowed aircraft to land in any direction.

During the early 1920s, pilots and airport managers considered the use of turf as an airport surfacing material beneficial. Because aircraft did not have brakes, grass provided friction that reduced stopping distance. Airplanes had tail skids to slow them, which made the self-repairing feature of good turf vital. Grass fields also kept down dust. Aviators believed that blowing dust and dirt would keep the public away from airfields, making aviation commercially unappealing. Dust also entered aircraft engines through carburetors, causing rapid wearing of cylinder surfaces and bearings, reducing aircraft reliability, safety, and profitability. Finally, pilots also believed that turf was softer than hard surfaces and produced less stress on their aircraft's landing gear, another seeming advantage of the prepared grass field.

There was, therefore, an entire industry devoted to the maintenance of grass airfields. One can find numerous articles and advertisements in the aviation magazines of the 1920s and 1930s extolling these virtues and instructing airport owners and managers in the proper methods of draining, grading, disking, seeding, and rolling grass fields. The image of the efficacy of turf as a field covering was reinforced with references to the great "model" aerodromes of Europe, Croydon in London and Templehof serving Berlin.[6] Both retained grass landing and take-off areas (and, in a departure from U.S. practice, both were roughly circular) but had surfaced taxi and parking areas.

Yet the all-over field was obsolete in American aviation by the end of the 1920s. Although military airfields and those airfields not on the major commercial routes remained turf all-over fields into the late 1930s, airfields serving major U.S. cities adopted various forms of surfaced runway in the late 1920s. What brought about this divergence of commercial and military practice?

Three intertwined factors influenced the departure of civil practice from military. The first was aircraft design and, in particular, the tire pressures chosen for aircraft use. The second was the nature of civil aviation as a business, particularly in its demand for regularity. Finally, the geographic diversity of the nascent commercial industry imposed soil and climatological conditions that, when combined with the above factors, drove airport managers to adopt surfaced runways rapidly during the late 1920s.

Aircraft designers have to consider a number of external factors in the making of their creations. One of the least sexy, little regarded, yet very important such factors is the surface from which the craft is expected to operate. The most obvious difference results from a choice of water rather than land as the operating surface. Planes designed for water used floats, boatlike hulls, or both, while land planes had wheels (for ground) or skis (for snow and ice). The Wrights had initially used skids, but builders abandoned them rapidly as they began to design aircraft too heavy to be simply picked up and carried off the field, as the Wright Flyers were.

The apparent lack of options for landing on the ground (wheels or wheels) might suggest that designers had little choice, but in fact there is wide diversity in wheels, particularly in tires. Bicycles, for example, rely upon different tire pressures for different surfaces: mountain bike tires are typically thirty-five pounds per square inch (psi) for mud and soft dirt to sixty psi (for hard ground), while common road bike tires run as high as 110 psi. The tire pressure is, to within a few psi, the pressure a bicycle (or plane or car) exerts on the surface, and each sort of surface can bear a different load. A 1944 textbook on airport design reported that most soils could withstand between fifty and seventy psi once prepared by rolling and when dry, but moisture reduced that bearing capacity to as little as ten psi.[7] Tires with higher pressures than a particular soil can support dig into it, which is why road bikes tend to damage lawns (especially wet ones), and why dirt bike rallies inevitably trash the soft soils they are held in.

This is also one of the factors in the failure of turf airfields in the United States. A U.S. Army Air Service circular from 1922 lists the standard aircraft tire pressures in use at the time as fifty to sixty psi, and its authors demonstrated that those pressures translated into pressures of fifty-two to seventy-one psi exerted on the ground.[8] The 1944 text shows these pressures were acceptable, if marginal, for smooth, dry turf fields, but moisture meant a day (or more) of no flying lest the field be damaged.

There were two options available to engineers interested in solving this landing problem. One was to use lower pressure tires on aircraft, as the bicycle example suggests. Private aircraft owners often did just that, adopting low-pressure "balloon" tires (ten to twenty-five psi, in many cases) because such tires allowed the convenience of operating from relatively unprepared fields. That made the airfields they used less expensive to build and maintain.

For larger aircraft, however, the low-pressure tire approach rapidly became infeasible. In order to support an aircraft of the same weight, a low-pressure tire had to be much larger than a high-pressure one because the tire had to distribute the same load (the aircraft's weight) over a larger area. Increasing the plane's weight would result in even greater increases in size. Larger tires meant more drag on the aircraft in flight, while again increasing the aircraft's weight. Neither of these effects was desirable to commercial and military aircraft designers, for different reasons. Commercial operators were initially paid by weight for carrying the mail, and therefore minimizing aircraft weight in favor of greater mail capacity was of economically valuable to the airlines, as was reducing drag in favor of fuel economy. Military aircraft have always been designed for high performance, and increased drag and weight imposed performance penalties that were unacceptable to military customers. Retractable landing gear, introduced to commercial aircraft in the early 1930s, was one consequence of the desire to improve fuel economy and performance by reducing drag.[9] Once tires had to be stored inside the aircraft in flight, designers were under further pressure to reduce the size of tires, which again meant increasing tire pressure.

Nevertheless, aircraft designers were able to keep aircraft tire pressures within the fifty to seventy psi range through the mid-1930s (the Douglas DC-3 used sixty-three psi tires) in order to maintain aircraft compatibility with existing fields. After that, however, the massive increase in aircraft size made this impossible. The 1935 Boeing B-17, for example, weighed twice as much as the DC-3, so its tires would have been twice as wide as the sixteen-inch tires on the DC-3 at the same pressure. Storing a thirty-two inch wide tire was a great waste of space, and tire pressures therefore reached one hundred psi by the onset of World War II. The escalation stopped there only when designers adopted multitired landing gear.[10]

The commercial and military demand for higher performance aircraft of ever greater weight meant that the low-pressure tire solution to the landing field problem was no solution at all. At best, it was a stopgap measure until the other possible solution could be implemented. That measure was the adoption of artificially surfaced runways. Artificial surfaces promised to overcome the weather sensitivity of turf fields just as they had helped overcome the same problems for automobiles and bicycles. The need to overcome weather sensitivity, in turn, descended directly from the commercial aspects of aviation.

Airports are designed to deal primarily with aircraft traveling from place to place in the pursuit of commerce, just as the primary function of seaports is to deal with commercial ships hauling goods from place to place. City planners immediately grasped the commercial consequences of airports and deliberately chose sites based upon relative proximity to surface transportation. Since people and goods carried by aircraft had to get from the airport to their actual destination, architects and city planners focused on airports as efficient nodes of transfer.[11] Selection of airports as efficient and safe landing sites was distinctly secondary, as was the consideration of site suitability for proper turf growth.

The primacy of efficient location of the airport with respect to surface transportation in urban areas was enhanced by a second commercial influence on aviation. That is due to the simple reality that people (especially business people) only want to go where other people are. The United States was an urban nation already by 1930, and its population was rapidly becoming more centralized. In addition, the largest cities were within the snow-belt states. With those demographic and geographic realities, the Post Office established the transcontinental mail route across the northern states in the 1920s: Washington, New York City, Cleveland, Chicago, Salt Lake City, and San Francisco were the major stops. Of these, only Washington and San Francisco were relatively free of snow, and the drained swamp upon which Washington sat proved to be at least as troublesome for airport building. Rainfall along most of the route was substantial, and the soil tended to retain water, which was good for growing but bad for building load-bearing structures. The demographics of the United States demanded, in sum, that commercial operators, building upon the Post Office mail routes, operate in precisely those places that Benjamin Foulois had argued should be avoided by the army in establishing its airfields: the cold, wet, foggy, snowy North.[12]

Finally, commerce dictated regularity. Although the greater speed of aircraft made air travel appealing to the first brave passengers (almost always business-people), more important for the issue of runway development was business travelers' demand for scheduled service. Airplanes are not faster than trains if they have substantial weather-induced delays, such as an airfield turned to mud by a sudden storm or the spring thaw. Diversion of incoming planes to other fields from similar causes was equally unwelcome to business travelers. This, too, was reinforced by the Post Office, which insisted its airmail contractors deliver the mail daily, regardless of weather or airport conditions. Private, and indeed military, fliers could and did wait for airfields to dry out enough after a storm to resume flying without damaging the turf surfaces, a practice common even during World War I.[13] The mail could not wait. Airfields whose owners wished them to be *airports* had to ensure continuous availability of their fields to attract a mail contract, and therefore an airline.

These three demands of commerce conspired to place major commercial airports in what were relatively poor sites in terms of their suitability as landing fields. The meteorological conditions of these areas, combined with soil conditions at specific sites, brought about the rapid demise of grass as a commercial airport surface technology in the United States.

As early as 1923, the problem of maintaining turf fields for aviation was recognized by construction engineer Archibald Black, who stated that the need for special runways had grown out of the breakdown of turf under regular use. This breakdown was accelerated by the "softening" of the turf that occurred during "certain periods of the year." Softening, caused by higher soil moisture content, meant that aircraft tires did more damage. This was especially true where the soil froze over the winter, which prevented drainage of surface water for several weeks during the spring. Freezing areas with soils of high organic content were also subject to "lensing," which created zones of weakness in the soil.[14] Black noted that there was a trend toward specially prepared runways, and he predicted that the tendency would increase. Black did not refer to runways in our modern conception, however.

Black specifically identified several possible runway surfaces that, while not hard in the sense that concrete is hard, were certainly considerably harder than turf. Based upon standard road construction practices, Black reviewed the possibilities of using gravel, crushed stone, cinder, and asphalted, or oiled, runways. Black's articles were reinforced by a 1929 textbook he published on airport engineering.[15] In it he focused on the similarities between the needs of automobiles for a good road surface and the needs of aircraft for take-off and landing surfaces, finding that the two machines had essentially the same requirements. He prescribed working crushed stone or gravel into the ground through repeated watering and rolling, and crowning this with oiled cinders, sand, or smaller gravel, depending on the local availability of these materials. These recommendations were based upon existing road construction practice, as he readily admitted.

Black's review of these surfacing methods suggests that he expected improved drainage to be insufficient to keep turf fields in reliable service at many sites. His analysis was supported in many other articles published during the later 1920s. Quincy Campbell, an Air Service reserve pilot, stated that sod was initially the best choice for fields, but increased traffic had made them unmaintainable. Tail skids cut up the sod in "soft" weather, and ruts developed from wheels. Frequent rerolling to eliminate ruts further damaged the sod, resulting in dust and mud.[16]

Campbell did not discuss the cost involved in turf maintenance, but it was clearly substantial. Removing fields from use for rerolling and reseeding meant less revenue for airport operators as well as increased direct costs. The cost to aircraft owners of poorly maintained fields is difficult to assess quantitatively, as no statistics breaking out surface-caused accidents from other types exist. Qualitative examination is easy, however. *Airports* contained an "Airports in Pictures" section often devoted to photographs of aircraft wrecked by poor field conditions. Although these accidents rarely appear to have been fatal, they represented an unacceptable financial risk to the infant airlines created in the wake of the Post Office's 1925 decision to privatize the airmail. Maintenance of turf was clearly a substantial burden that at least some airport operators were either unwilling, or unable, to shoulder.

The airlines, the Post Office, and the Department of Commerce drafted an air-

port rating standard in 1926 that substantially influenced a wave of airport building that began after privatization of the airmail. The standard was not mandatory,
as the newly formed Aeronautics Branch of the Commerce Department had no
legal authority over airports. Nonetheless, the rating standards provided a framework for comparison and certification of airports.[17]

Cities often went well beyond the standard and certainly did so with runways.
The standard only required year-round serviceability and a minimum width and
length. It did not require surfacing of any kind, stipulating vaguely that fields be
smoothed to the point that a passenger car could drive over it at thirty mph without "undue discomfort" to the occupants.[18] Yet cities across the country built surfaced runways. A few examples will suffice to show the reasons behind cities' decisions to adopt runways over turf.

In 1929, the airport superintendent for Columbus, Ohio, reported to *Airports*
magazine that he had chosen paved runways over turf due to airline demands for
regularity of service and the high volume of traffic his city hoped to attract. Accordingly, his runways were built with a five-inch concrete base with a 1.5 inch asphalt crown. This decision drew an attack from a soil engineer named Wendell
Miller, who argued that competent soil engineers could have built an "all over"
turf field suitable for year-round use for half the cost. "Soil technology," he wrote,
"as a definite engineering science, capable of lessening the cost of airports, has
not yet been recognized by the officials of air transport companies who have had
the direction of airport construction policies."[19]

At issue here was the problem of drainage. Since aircraft certainly could use a
turf field in dry weather, airline demands for regularity required some means of
maintaining a dry field in wet weather. Miller believed that a properly designed
drainage system could achieve this. Unfortunately, neither article addresses the
system installed at Columbus (all airports, turf or not, had drainage systems of
one sort or another), but details on other wet climate fields are available, and using these we can at least qualitatively explore the limits of drainage.

The supervisor of public works for Rochester, New York, described his city's
choice of flying field and the work that had to be done to make it useable. Stating
his preference for turf fields, he wrote that the field selected by the city, located in
a bend of the Genesee River so that it could serve seaplanes and float planes too,
tended to be waterlogged near the center, and the heavy, clay-rich soil exacerbated
the problem. He had laid 7,500 lineal feet of drain tiling to remove the excess water from where the runways were to be built. It was not enough piping to provide
a firm surface for an all-over field, however. It was simply enough to protect the
runways from being undermined. At Rochester, the pipes were placed under the

runway centers, further minimizing construction costs. The runways themselves were built of 17,000 tons of stone brought in by team and wagon. The stone was worked into the soil, and crowned with several inches of cinders bound with asphaltic oil to make the runways waterproof and provide a resilient surface capable of withstanding tail skids.[20] Later, the cinders were replaced with asphalt because cinders that worked loose tended to damage propellers. Rochester's superintendent chose what he thought was the least expensive approach to building an airport, with limited drainage intended specifically to protect the runways.

Similar reports of airport construction abound. Iowa City, Iowa, laid 150 by 2,800 foot runways consisting of an eight-inch crushed limestone base, packed hard and rolled. Drain tiling was again installed to protect the runways, due to an "excess of moisture during winter months, freezing and thawing of the ground, [and] the heavy black soil." St. Paul, Minnesota, one of the first additions to the transcontinental mail route, spent $295,000 in building its municipal airport in 1926 and floated an additional half-million dollars worth of bonds by 1929. Like Rochester's original airport, St. Paul's was built on low land next to a river (the Mississippi). Here, the problem was drainage and snow removal. Photographs show that plowing the runways for the mail planes coming from Chicago (which had snow far less often, and therefore mail planes could not fly in and out with skis) severely damaged the surface. This meant an expensive reconstruction job each spring, until the airport replaced turf runways with hard surfacing.[21]

Airport managers chose a combination of drainage and runways to reduce the construction costs of their fields. Runways, like roads, required a firm, dry subsurface to prevent cracking of the pavement, and airplanes needed a firm, dry surface to prevent tire rutting and bogging down. Managers could not avoid drainage entirely, but the use of runways reduced dramatically the areas that had to be drained. Drain piping had to be buried between four and twelve feet down, with wetter areas requiring shallower piping. Shallower piping, in turn, drained a smaller surface area, which meant fields in wet areas required a great deal more piping. An all-over field in a wet area, such as Rochester, meant an extensive subsurface drainage system. Having to drain only runway areas seemed a cost-effective solution, especially in the light of the army's well-known drainage problems with two of its major fields during the 1920s.

Perhaps the best evidence of the inability of drainage alone to solve the problem of boggy airfields at a reasonable cost comes from the army's struggles at Selfridge Field near Detroit and the original Bolling Field near Washington, DC. Both had been chosen for their proximity to important cities during World War I. Selfridge's location was chosen to ease the testing of aircraft that the automobile

industry was expected to produce for the war effort; Bolling's location was chosen for easy access to the army's bureaucratic center. Both had terrible drainage.

Selfridge was the second army field to receive a surfaced runway after McCook
Field in Ohio. Despite the army's continued adherence to all-over fields, it was
forced by conditions at low-lying Selfridge to install runways after sixty-three
miles of drainage tiling failed to produce a useable year-round field.[22] The efforts
of Rochester, in laying a mere 1.5 miles of tiling, pale in comparison.

The army's experience with Bolling Field was worse still. Originally, it was located on the Anacostia River, on hydraulic fill dredged from the river. The Inte-
rior Department had intended to make the area a park until the army conducted
a campaign in Congress to get possession of the field. Because the field was only
a few inches above the river, it was prone to repeated flooding. Consisting of the
extremely fine stuff that typically makes up river bottoms, the soil was largely im-
permeable and did not drain. Drainage piping proved useless, as the ground a
few inches down was fully saturated. After an embarrassing cancellation of a major exercise in 1927, the army began a new campaign to move the entire field to
a nearby site, which Congress approved in 1929.[23] In the meantime, the army
built a runway on the old site, in order to get some use out of the field. These well-
publicized failures no doubt impressed upon airline and airport managers that
even massive drainage systems might be insufficient when faced with low-lying
fields in wet climates and no doubt influenced their decisions to reject soil engineers' claims for all-over turf fields in favor of runways.

Bolling and Selfridge were poor choices from the standpoint of flyability, and
the army paid a high price for its choices. But just as the army chose Selfridge
and Bolling Fields without much attention to geography, cities and city planners
were more concerned with placing airports near other forms of transportation
than with choosing high-quality flying environments. The scarcity of large, un-
developed tracts of land near surface transportation in urban areas severely limited the choices of sites available to cities. Some cities chose to make new land,
as the number of airports built on fill attest: San Diego's Lindbergh Field, Boston's
Municipal Airport, New York's La Guardia, and San Francisco's Municipal were
all built during this period at least partly on fill. Fill, especially hydraulic fill as the
Bolling example suggests, seldom resulted in high-quality, naturally drained land,
thus leading to the use of runways. The claims of soil engineers notwithstanding,
it seems clear that the efficacy of drainage had its limits, and many municipal airports exceeded them. This problem, combined with the unwillingness of military
and commercial organizations to accept the performance limitations imposed by
low-pressure tires, meant that runways were the only realistic solution. Con-

fronted with snow and ice removal in the north, water nearly everywhere, and the demands placed by airlines on regularity, safety, and efficient transshipment of passengers and mail, the technology of the all-over turf field became obsolete in the United States during the 1920s.

MAKING A TECHNOLOGY FIT:
EVOLUTION OF THE LEADER CABLE SYSTEM

The obsolescence of grass fields in turn impacted blind landing technologies intended for them. The United States, the United Kingdom, and France had all experimented with balloons, acoustic location, and "leader cables" in their attempts to solve the blind landing problem during the Great War. After the war, primary interest in France and Great Britain focused on leader cables. In France, the system was known as the "Loth system" after its inventor William Loth; in Great Britain and the United States, it was simply called the leader cable system. A number of variations were developed, and I use the term *leader cable system* to refer to the class of systems in general and designate specific incarnations by the name of the inventor.

Charles Stevenson of the Royal Society of Edinburgh first proposed the leader cable system's basic idea in 1893. He described the testing of a single cable laid along the sea floor to see if induction through sea water could happen without the use of parallel wires. Stevenson undertook this experiment after he had the idea that such a cable could be of great navigational assistance to ships. He laid a single cable on the floor of a lake and fed it with eighty volts to produce an electromagnetic field. The combination of an uninsulated copper wire coil hung at each end of the boat, connected with a wire hooked to a telephone receiver, acted as a detector. Using the telephone, an operator could hear a tone that increased in intensity as the boat neared the cable, as long as the wire in the boat remained roughly perpendicular to the cable. Once the boat was centered over the cable, the tone stopped. The author presented no scientific analysis of the device's function, and how, exactly, the leader cable system worked remained subject to debate in France through the 1930s. Lack of scientific understanding of a useful new device rarely stands in the way of its adoption, however, and the German navy put a version of this idea into service before the Great War. The British Royal Navy became interested in leader cables after finding out about them from captured German sailors and installed an eighteen-mile cable run into Portsmouth. It abandoned the idea due to the high cost of maintaining under water cable, particularly since ships' anchors tended to hook and cut it.[24] The Royal Navy's rapid abandonment of the nautical leader cable anticipated some of the reasons for the aeronautical version's failure.

France took the leader cable idea the furthest after the war, developing a guidance version for point-to-point aerial navigation and a landing-aid version. The French government voted to build guide cables from Paris to London, Paris to Brussels, Paris to Strasbourg, and from Paris to Versailles in 1922, and considered building one across the Sahara desert. The British Parliament approved funding for the section of the Paris to London route on British soil in 1923. The records available in the United States suggest that little was actually built, but they are hardly definitive. The one published work on air navigation before World War II does not discuss leader cables at all and indicates that aerial navigation in Europe was carried out through the use of radio direction-finding stations. However, a series of articles in the French aeronautical press during the 1930s debates the merits of the leader cable system, suggesting but not confirming that at least parts of the planned system were built.[25] The records related to the landing-aid version are fortunately somewhat better.

The first published reference to any of these aircraft systems is Loth's "On the Problem of Guiding Aircraft in a Fog or by Night When There Is No Visibility." It was presented to l'Academie des Sciences in December 1921, and it described the system for overland navigation and blind landing of aircraft based on the generation of an artificial electromagnetic field. He had been asked to develop the system by the undersecretary of state for aeronautics and had built his prototype at Villacoublay. To provide guidance for aircraft trying to land, his system passed an alternating current of six hundred cycles per second through a loop of cable surrounding the landing field. This produced a varying electromagnetic field that a properly equipped aircraft could detect.

An aircraft's equipment consisted of three sets of two frames, or loops, oriented along the aircraft's three axes. Passing any wire loop through a magnetic field perpendicular to the field's lines of force induces a current in that wire, and to detect the current induced in the wire frames, the system employed an audio amplifier. Two sets of the frames were vertical, with one aligned with the longitudinal axis of the aircraft, and the other perpendicular to it. The third frame was horizontal. The longitudinal frame received best when parallel to the guiding wire, while the transverse frame received best when perpendicular to the wire. They could be connected to a goniometer to display the aircraft's inclination to the guide wire, while connecting the longitudinal and transverse frames indicated whether the aircraft was right or left of the guide. Finally, connecting the transverse and horizontal frames showed vertical position with respect to the cable. It thus provided three-dimensional navigation information. Early versions of this system relied upon a pilot's hearing to interpret the system's audio signals. Later versions employed visual indicators on the pilot's instrument panel.[26]

French engineers published a series of articles in the journal *l'Aéronautique* during the 1930s which explained the operation of the system and described the results of tests conducted at the aerodromes of Villacoublay and le Bourget. Unfortunately, those articles argue the scientific rationale of the system and give no insight as to how effective it was or how widespread its use became in France. Loth also founded a company to develop and manufacture his system, the *Société Industrielle des Procédés Loth*, and even an American subsidiary, the American Loth Company.[27]

Loth's system does not appear to have been placed in either commercial or military service. Although no source provides a reason, Christienne's history of French military aviation provides one clue. The French aeronautics establishment was fiscally poor and organized to create prototypes but not actual procurement and production. The result was the rapid obsolescence of French aircraft and aviation infrastructure. The death of the Count de la Vaux, who was killed along with his pilot when his plane collided with a landing cable installed at Vaux sur Seine, did not help the system's chances.[28]

Across the Channel, British investigators working at the Electrical Research Section of the Royal Aircraft Establishment (RAE) had begun exploring a similar application of electromagnetic principles to solving the blind landing problem some time before 1920. Unfortunately, security concerns prevented publication of their work before 1926. Led first by one Major McAlpine, and after 1923 by Flight Lt. H. Cooch, the RAE's team designed a blind landing system similar in principle, but significantly different in detail, to the Loth leader cable system. Three key differences were the use of a racetrack shape for the cable, with one of the straight sides marking the landing area; the use of thirty-four Hz current in the ground cable; and the use of a frequency reverser in the plane to convert the received thirty-four Hz energy to direct current.[29]

The use of a mechanical frequency reverser was the most significant difference in the British system. The French had relied upon the generation of a weak electromagnetic field by the cable and used a vacuum-tube-based audio frequency amplifier in the aircraft to amplify the signal. The British chose to generate a much stronger electromagnetic field on the ground, and because direct-current reading instruments were more sensitive than alternating-current instruments, convert the received alternating current energy in the plane to direct current via a mechanical-relay-based device. Because mechanical relays could not operate very rapidly, the frequency of the generated electromagnetic field had to be kept low. This created two problems for the British. First, as was becoming well known, low-frequency signals had unpredictable propagation characteristics. Worse, the Electricity Commissioners had recently chosen fifty Hz as the standard frequency for England's electric train system.[30] The nearness of the two frequencies, coupled with the much higher-intensity electromagnetic fields generated by the very high current use of the trains, resulted in significant interference. Although the system tested successfully in a small-scale laboratory setting, when fully installed at Farnborough, it did not do so well.

Cooch presented his findings to the Royal Aeronautical Society on February 19, 1926, and the audience's response makes clear that it was well aware of the frequency problem. They were also aware of the French Loth System and of the theoretical benefits that they could gain through its use of higher frequencies. Higher frequencies, however, required the use of higher speed switches of some sort or other, and only the vacuum tube, or as the British called them, thermionic valves, were suitable for use either as high-speed switches, or as amplifiers. Neither Cooch nor his audience that evening believed that thermionic valves were reliable enough or, apparently, that they could be made reliable enough for use in a safety-related system.[31]

Therefore, by 1930, the RAE had entirely abandoned the leader cable system. Instead, it began pursuing an idea that air forces had tried unsuccessfully during World War I: the tethering of a balloon above the fog where pilots could see it. A pilot seeing the balloon maneuvered as close to it as he could and then simply descended at a constant rate along the proper compass bearing until the plane hit the ground. The RAE realized that such a landing would be at best frightening to passengers, and so it devised a "ground proximeter," consisting of a weight hanging from thirteen feet of linen thread below the wheels. When the weight hit the ground, the tension on the thread was released, causing a light to flash on the instrument panel. This was to give the pilot enough time to pull the nose of the plane up to reduce the shock of the landing. A strengthened undercarriage was still required.[32]

The ground proximeter did nothing to resolve the problems that World War I fliers had encountered with the balloon method. The altitude of fog banks changed over time, often obscuring the balloon, and the inaccuracies of barometric altimeters meant that a constant rate descent was often not constant at all, which occasionally led to reaching the ground outside the boundaries of the aerodrome. Pilots tended to be uncomfortable about such possibilities and in the

United States rejected another system that also relied upon the barometric altimeter.[33]

Finally, as was pointed out by F. Tymms of the Directorate of Civil Aviation during a discussion of the balloon test results, "However valuable this result might be, it could not be considered a satisfactory solution for the needs of air transport, whose successful development demands *regularity*. Further, if pilots were merely to use the system when accidentally caught in fog, they would never attain confidence in it."[34]

The emergency nature of the balloon system ensured that it did not appeal to commercial aviation interests. Because the balloons had to be deployed whenever poor conditions occurred, they were not always available. Pilots would not be able to practice using them during good weather, and practice was the only way to develop pilots' confidence in their ability to use the system safely. An emergency use only system was not what commercial airlines sought, and it is surprising that the RAE seriously considered the balloon landing idea at all. The British abandoned the leader cable system, which could have been a part of everyday flying, for an emergency system that could never be routine. Rejecting thermionic valves thus led them to a technological dead end that is all the more surprising given that wireless, which relied upon tubes, was already in widespread use for navigation in Europe and the Empire.

In Europe, therefore, no single cause for the nonadoption of leader-cable-based systems for landing aircraft exists. Instead, each country had a unique experience. In France, an idiosyncratic state policy is the likely cause. In Great Britain, a conscious choice against vacuum tubes provoked abandonment and replacement by what was an even less suitable procedure. Germany, whose aviation activities were severely limited by treaty restrictions, did not conduct blind landing research until 1931, when Telefunken, Lorenz A.G., and Deutschen Versuchsanstalt für Luftfahrt, the state-supported aeronautics research laboratory, embarked upon a radio-based development program based loosely on a U.S. Bureau of Standards' idea.

Although the leader-cable-based systems failed in Europe, they nonetheless fit within the surrounding technologies of airplane and landing field. They did not produce great precision, but none was necessary to place an aircraft somewhere within the confines of a large aerodrome. They were thus suited to open-field landings by aircraft nimble enough to negotiate a relatively short approach, which were all that existed in Europe throughout the mid-1930s. They did not fit within the technological context of American aviation, however. The U.S. need

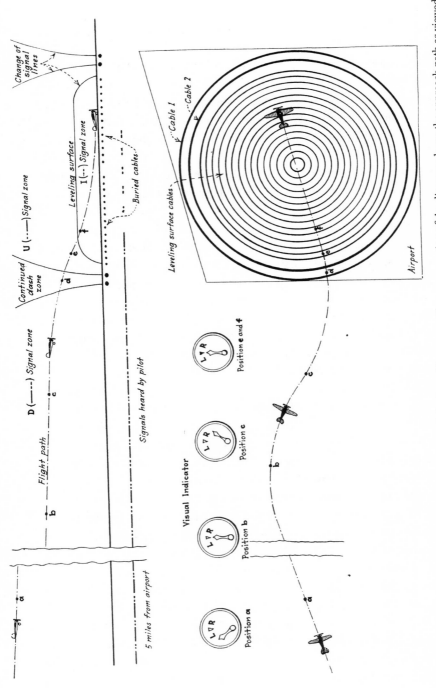

Figure 2.1. The American Loth Company's version of the Loth leader cable system. The upper portion of the diagram shows the approach path as viewed from the ground, with the Morse code signals displayed in the different approach zones. The "leveling surface" is simply the magnetic field intended to indicate to the pilot that it was safe to level out and land. The lower portion of the diagram shows the bird's-eye view of an approach, keyed to the visual indications that the system provided to a pilot. Notice the lack of anything resembling runways on this figure. *Aviation* (December 1931): 199.

for greater precision was due to the adoption of runways in North America more than a decade before their widespread use in Europe. To provide enough precision for runway landings, the leader cable idea had to be modified.

During the war, the United States had also explored a blind landing system based upon the leader cable principle. On Armistice Day 1918, the U.S. Bureau of Standards tested a prototype consisting of a "transmitter" composed of a 160-foot coil of copper wire, which the bureau's researchers wrapped around the roof of the Radio Building, powered by a 500-watt generator. Wire coils wrapped around the lower wings of a JN-4 "Jenny" biplane connected to an audio frequency amplifier served as a detector.[35] Although the pilot was able to hear the signal the system produced, he did not attempt to land on the building's roof using it, likely because the roof was too small. Several more months of testing at the College Park, Maryland, airfield followed, but postwar austerity measures put an end to the project. The Post Office, which assumed airmail responsibilities from the Army Air Service after the war, considered navigation between airfields a more important problem and put its limited funding into the construction of lighted airways. No further organized research into blind landings was done in the United States until 1927, when the Bureau of Standards was given the task by then secretary of commerce Herbert Hoover. In Britain and France, however, blind landing efforts continued under government auspices and focused on the leader cable approach.

In the United States, proponents of leader-cable-type systems developed three different variants in an attempt to make the idea fit into a different context. The first was a product of the American Loth Company and built upon Loth's work in France. Beginning in 1930, the Company modified his original formulation to replace the single, aboveground loop with a series of concentric underground cables, which may have been a response to Count de la Vaux's death (Figure 2.1). As a plane flew toward this installation, its pilot first heard a Morse code *D* signal until the craft crossed the field boundary, where *D* was replaced with a continuous tone. Inside the field, the system provided a Morse code *U* until the aircraft was twenty feet above the surface, when the small inner cable's *I* signal overwhelmed the *U*. That was the signal to land.[36]

This ingenious arrangement met with no greater success than its parent system. No one ever adopted it. By 1932, all major U.S. airports had adopted runways, and as the geometry of the system makes very clear, it could not serve a runway-based airfield. Like its European forbears, it relied upon the provision of an open landing area. The circular form allowed landing from any direction, which a runway-based airport could not permit. Although it might have served the aerodrome-like U.S. Army fields, the army emphasized the need for porta-bility in any blind landing equipment throughout the 1930s, which the Loth sys-tem certainly was not. The American Loth Company's system remained too much like its European forebear to fit the U.S. technological context. Instead, inventors had to modify the leader cable arrangement to suit U.S. airfields.

While the American Loth Company worked on its circular system, Earl C. Hanson, who had installed a leader cable for ships in New York harbor, designed a system that could serve runway-based fields. This was installed and tested at the Lansing, Illinois, Ford Airfield, which had opened in 1928 with the first concrete runways in the United States (Figure 2.2). It relied upon two single-turn cables extending nearly four thousand feet beyond the airfield boundaries and aligned along the approach path. By exciting both cables with a 1,000 Hz frequency and using switches to alternately open and close the circuits, a "dot" pattern was ra-diated in one cable and a "dash" pattern in the other. A pilot flying precisely halfway between the two cables heard a continuous tone. Further, a "gun coil," which produced a loud tone when the plane passed directly over it, informed the pilot when the plane had arrived inside the airfield boundaries. An inductive al-timeter provided altitude indication. Hanson's successful demonstration in 1930 drew congratulatory telegrams from Lee De Forest and Capt. Stanford C. Hooper of the Navy Department's radio section, who took interest in the basic design of the system for its applicability to airships.[37]

Hanson's work seems to have inspired Hooper to support further develop-ment of the basic idea, which was carried out by a navy radio engineer working for Hooper's Radio Section of the Bureau of Engineering. Edward Dingley de-vised the system shown in Figure 2.3 in the mid-1930s. The system fed alternat-ing current through the cable arrangement shown in the diagram, from the gen-erator labeled "3." The cable set defined by points 36-37-3 was the main loop, which generated a powerful electromagnetic field. Each of the smaller loops (for example, 38-39-3) was fed with current of the opposite phase to that in the main loop, establishing an opposing electromagnetic field. This resulted in destructive interference between the two fields. By carefully choosing the currents in each ca-ble, engineers could establish an electromagnetic field whose intensity decreased as an aircraft approached the runway. Using a radio receiver designed to detect a line of constant intensity, the pilot could follow what would appear as a straight glide path to the runway. The aircraft installation consisted of the same wire frames, in the same orientation, as those originally proposed by Loth and were connected to a cross-pointer instrument much like that proposed by the Bureau of Standards in 1931.

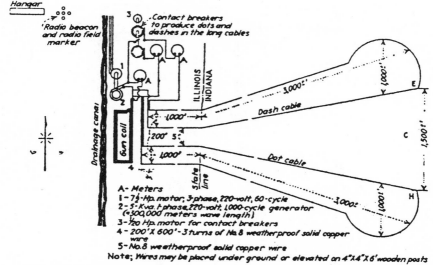

Diagram showing the circuits. Note form of loop employed.

Figure 2.2. Hanson's modified leader cable system. The diagram is clearly not to scale, but it demonstrates how this system could serve runway-based airfields. The center "on course zone" is 200 ft wide just before crossing the airport boundary, which was a common runway width. The approach length is a bit short even for 1930 aircraft. Other systems designed between 1929 and 1933 generally supported a one- to two-mile approach. Since commercial aircraft of the day landed at about 60 mph, Hanson's approach provided about half a minute of guidance before landing. Note that the system provided no altitude indication, a problem that caused pilots to reject other landing aids during the 1930s. *Aviation* (22 February 1930): 402.

Dingley advertised its chief advantages as its independence of radio frequencies and the consequent atmospheric and ground distortions. It did not employ vacuum tubes, and Dingley argued that this made its maintenance costs negligible.[38] The electromagnetic field provided lateral and vertical guidance, so a separate localizer was unnecessary, as were marker beacons. Finally, since the system did not require antennae on the ground, or perhaps more properly, the antennae were underground, it presented no obstructions on the field.

To test Dingley's vision of a landing system, the navy built a prototype within the confines of its airship base at Lakehurst, New Jersey. That forced a short approach and an accordingly steep descent, which was fine for lighter-than-air craft but impossible for airplanes. The navy considered lengthening the approach path enough for airplane use during 1940, until the Lakehurst base commander

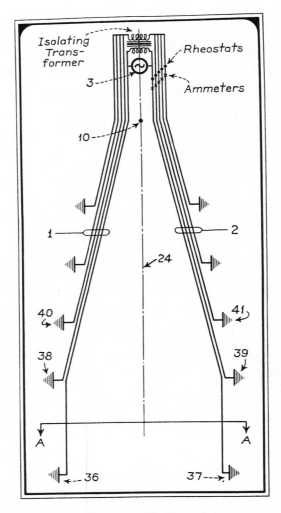

Figure 2.3. The Dingley leader cable system. Like Hanson's it was supposed to guide planes down a narrowing funnel of electromagnetic energy. Unlike Hanson's, it provided a "glide path" through the use of successive additions of oppositely polarized electromagnetic fields. Destructive interference between the main field and the succession of smaller fields resulted in a decreasing field intensity which appeared to the pilot as a glide path. By 1938, the army was demanding approaches of at least ten miles, and preferably fifteen, for its aircraft. For this system, the cable array would have had to be extended the entire approach distance. Edward Dingley Jr., "An Instrument Landing System," *Communications* (June 1938): 7.

pointed out that lengthening the approach meant having to secure easements on land well outside the base. In Lakehurst's case, that meant having to get easements on cranberry bogs, which he expected to be expensive and an installation nightmare. It was briefly examined by Vannevar Bush's National Academy of Sciences committee on instrument landing aids in 1939 and by the CAA at the same time, but both considered radio a better solution, most likely due to the high costs involved in building and maintaining what would have been ten-mile-long cables in order to guide the much larger, faster aircraft of the late 1930s to safety. Admiral Stark, the Chief of Naval Operations, finally ordered the project's cancellation in September 1941.[39]

CONCLUSION

One clear factor that emerges in discussion of landing aids and landing areas is that of the environment in which the technologies of aviation had to work. Turf fields and leader cables were not the only technologies affected by the physical environment of aviation, and later chapters show similar problems with other technologies. Technologies designed without explicit recognition of these environmental constraints had severe problems, and the leader cable designers were not the only innovators who failed to overcome them.

The environmental problems encountered by aviation technologies cannot be separated from economic factors, however. Cities could have, and in retrospect perhaps should have, put their airports inland, away from water problems for surfaces and landing aids, as well as navigation hazards imposed by urban structures. They did not do so, due to city planners' emphasis on airports as nodes of transfer, which reflected their primary function as spaces for commerce.

Leader cables were therefore made obsolete in the United States by the adoption of runways driven by the intertwined demands of commerce and geography. Although various inventors attempted to make the system fit the new technological context, the result was systems that fit just as poorly as their predecessors.

The intertwined demands of commerce and geography thus effectively doomed the leader cable to the same fate they had spelled out for the turf field, relegating both to the technological scrap heap. But the same airline demand for regularity which had played such an important role in creating runways, and which still regularly ran afoul of bad weather in the form of poor visibility, remained unsatisfied.

While leader cable inventors were tinkering with their systems, therefore, the airlines put pressure on the Post Office, which paid them to carry the mail, and the Department of Commerce, which maintained the federal airways for them, to find a solution to the blind landing problem. The government's proposal had nothing to do with leader cables. Instead, the commerce department turned to the only technical development group available to the civil half of the U.S. government: the National Bureau of Standards, which possessed a radio research laboratory.

CHAPTER THREE

Radio Blind Flying

Two government organizations played key roles in the late 1920s and early 1930s blind landing research, the U.S. National Bureau of Standards (NBS) and the Aeronautics Branch of the Department of Commerce. The U.S. Congress had established the NBS in 1901 to devise and maintain a uniform system of weights and measures for the nation. That mission included maintenance of frequency standards for radio transmission, and the bureau established a radio research laboratory in 1913 to carry out that responsibility.[1] Researchers at that laboratory created the basic model on which the current instrument landing system is based, the now familiar marker beacons, localizer, and glide path.

While the Bureau of Standards' authority was well established, the Aeronautics Branch was a new agency when blind landing work began. Civil aviation in the United States had been unregulated before 1926, with the exception of the Post Office's mail service. The postmaster general's decision to privatize airmail carriage in 1925 resulted in a new commercial airline industry and its almost instantaneous demands for federal regulation. Airline pioneers feared that each state would impose its own regulations in the absence of a federal oversight, resulting in chaotic operating conditions and consequently little chance of profitability. The airlines prevailed upon Herbert Hoover, the secretary of commerce, who helped draft and push through Congress legislation creating the Aeronautics Branch in 1926. That organization, in turn, was reorganized in 1934 into the Bureau of Air Commerce, still within the Commerce Department.[2]

Hoover had also arranged for a joint program between the new Aeronautics Branch, the Bureau of Standards' radio lab, and the Bureau of Lighthouses to replace the giant searchlights that the Post Office had used to light the federal air-ways with radio ranges that would better facilitate blind flying. The radio ranges were an outgrowth of an army project at Wright Field. They broadcast signals that pilots could hear in their radio headphones. Using the signals, pilots would be able to fly blind from range to range across the country once all the ranges were in place. Beginning in June 1926, NBS engineers had designed these ranges with the Aeronautics Branch's support, and to them also fell the task of producing a blind landing system. Unsurprisingly, the Radio Section chose to base the blind landing system on the range design.

The ranges' instability made the choice problematic. Pilots could not depend upon them to maintain a consistent course, and they dubbed the ranges, not en-tirely humorously, "rotating ranges." The tendency of these ranges to vary was dangerous but rarely fatal. Most of the time, the variations amounted to a few de-grees and were not enough to lead pilots into danger. Variation of a few degrees in a blind landing system, however, proved intolerably dangerous. To land safely under blind conditions on a runway, aircraft had to be aligned within fractions of a degree to the runway centerline, and stability was therefore vital. The Bureau of Standards sentenced engineers to decades of work on this instability problem by choosing radio as the basis of a blind landing system.

The instability descended from the use of low-frequency radio signals, in the 100 to 300 kilohertz (KHz) band. This was "best practice" in the 1920s. Higher frequencies were known but could only be broadcast at relatively low power. The lower-frequency AM stations were much more likely to suffer from environ-mental interference than were the higher-frequency FM stations. In this discus-sion, "environmental" conditions refers both to those interferences deriving from the "natural" environment, such as storms, and those extending from the "built" environment, like bridges, power lines, and buildings. Interference effects get worse as one moves down the radio frequency spectrum, and the radio ranges in use during the late 1920s and 1930s broadcast on frequencies far below those used in modern AM radio broadcasting.

At root, the problem blind landing pioneers faced was that radio waves prop-agate through the environment and are affected by it. Because "first" nature, the nonhuman environment, and "second" nature, the built environment, are in-consistent, radio propagation through them is as well. Fixing that instability problem meant finding a way around nature's inconsistency. Pavement had been the answer to the inconsistency of the Earth's physical surface, giving planes a smooth, flat, hard place to land; unfortunately, consistency in radio-based blind landing equipment proved much harder to achieve.

FLYING THE (UNSTABLE) BEAM: THE NATIONAL BUREAU
OF STANDARDS SYSTEM

The National Bureau of Standards had been asked by Secretary of Commerce Herbert Hoover in June 1926 to begin developing radio navigation beacons to replace the high-powered searchlights that had provided guidance along the Post Office's airmail routes. These beacons, which were called ranges even though they did not provide a pilot with a distance indication, provided four courses along which pilots could fly either toward, or away from, the ranges. The navigational "beams" were formed by overlapping two signals, one coded with a Morse code A and the other with a Morse code N, as Figure 3.1 shows. The airplane's receiver simply transmitted Morse code signals to the pilot's headphones, and pilots flew these ranges by trying to make the two signals merge into a single, continuous tone. When a pilot heard a continuous tone, the aircraft was in one of the "on course" zones shown in the figure.[3] This type of arrangement came to be called an equal signal, or equi-signal, transmitter.

Flying the radio ranges during the late 1920s was not a simple task. Staying on course meant pilots had to listen intently to the tones in the radio headphones for hours on end and use them to orient the aircraft in aural space. Because the ranges operated at low frequencies, they suffered from static interference that could overwhelm the real signals, and they also suffered from fading due to changes in the upper atmosphere. Further, the ranges had more than one task. Their signals were interrupted every thirty minutes or so for weather and traffic advisories, which were transmitted by Morse code. All of these issues made the ranges challenging to fly, and in bad weather their demands on a pilot's attention were exhausting. There are numerous anecdotal accounts of accidents caused by simple fatigue. Nonetheless, the ranges worked well enough to warrant deployment throughout the United States, and during World War II the army stripped many lesser-used air routes of their ranges so the equipment could be shipped overseas.

In internal NBS memoranda in 1928, Harry Diamond, an engineer in the radio section, proposed the use of a modified low-frequency four-course range (330 KHz) as a landing aid. This was to allow "blind" landings to be made by providing a pilot with information about the aircraft's lateral position with respect to the runway. The addition of two low-powered, high-frequency (97.3 MHz) beacons in line with the approach path, one at the airport boundary and one two miles away, would provide a pilot with two positive position "fixes."[4] A pilot following the di-

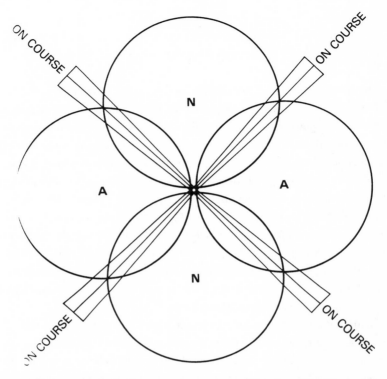

Figure 3.1. Radiation pattern of a four-course range, 1929. Pilots heard the Morse code A-N signals in their headphones, and maneuvered their aircraft so that the two signals merged to form a continuous tone, effectively orienting themselves in aural space. Ranges of this type were called aural ranges. Replacing the Morse code signals with two different modulation frequencies produced a signal that could be detected by an instrument. A range of that type was called a "visual range," and initially utilized the tuned reed indicator (see Fig. 3.2). Reprinted with permission from Nick Komons, *Bonfires to Beacons: Federal Civil Aviation Policy under the Air Commerce Act, 1926–1938* (Washington, DC: Smithsonian Institution, 1989), 157.

rectional beam (called a *localizer*) would pass over the outer beacon and begin descending at a constant rate. The inner marker, once heard, indicated that the plane was inside the airport boundaries and could safely land.

Doolittle's 1929 blind flying experiments at Mitchell Field were the first to employ the localizer idea. He borrowed from the army a Consolidated NY2 trainer and equipped it with special shock absorbers, a Sperry artificial horizon, a Sperry directional gyroscope, a specially designed Kollsman precision altimeter, and standard instruments. He also borrowed a low-frequency localizer from the NBS.

With this equipment, Doolittle was able to make the first deliberate blind landing on September 29, 1929.[5] He reported that over the several months of the experiments, several hundred blind landings were made with the NY2.

The localizer he used was a visual type, which was designed to operate a tuned reed indicator in the cockpit (Figure 3.2). The visual indicator was an innovation intended to replace the aural signals of the standard four-course range. It had a pair of metal reeds, each of which was mechanically tuned to resonate at the modulation frequency of one side of the equi-signal localizer. When the aircraft was on course, the two reeds appeared to be the same length. If the plane was to the right or left of course, the reed on that side would appear longer than the other, and the pilot would then turn toward the shorter of the two reeds.

Doolittle's landing procedure in the Guggenheim experiments was to fly toward the beacon maintaining one thousand feet altitude. When the aircraft passed directly over the beacon, the reeds stopped vibrating, signaling that the plane was over it. The box containing the reeds then had to be taken out of the receiver, turned over, and reinserted, while the aircraft flew about four miles. The pilot then made a 180-degree turn, turned over the reed box again, and descended to four hundred feet. Once the aircraft had passed over the beacon again, Doolit-

Figure 3.2. Tuned Reed indicator. This is a later version that fixed the problem of having to remove the reed container from its housing, flip it, and reinsert it during an approach. The "To" side was on top while flying toward a radio range, and the "From" side was rotated up when flying away from one. The two reeds are the white lines at the center of the indicator. From H. Diamond and F. Dunmore, "A Radio Beacon and Receiving System for Blind Landing of Aircraft," *Bureau of Standards Journal of Research RP 238* (1930): 905.

tle idled the engine, kept the wings level, and maintained a sixty-mile-per-hour airspeed as the aircraft descended to fifty feet. Once there, Doolittle opened the throttle again slightly to reduce his descent rate from one thousand feet per minute to six hundred, which was the most he could comfortably tolerate.[6] The aircraft's course was maintained with the directional gyroscope after passing over the beacon house. Doolittle then flew the plane into the ground. Because Mitchell Field was still a grass all-over field, it did not matter where on the airport surface he landed.

Doolittle readily admitted that this was an experimental system, and it never became more than that. The reed indicator, which was also used with the radio ranges along the federal airways system, was difficult to use, and the need to flip it over during an approach made it dangerous. Doolittle's precision altimeter, though novel in that it was synchronized by radio with a barometer on the ground, also gave pilots pause as a landing instrument because as a barometer, it merely read pressure above sea level and not above the ground. Perhaps more importantly, it did not show distance above obstacles on the ground, which were a significant hazard to fliers approaching an airfield. Barometric altimeters did not indicate what pilots called *absolute altitude,* which meant an aircraft's altitude "above terrain," including the ground and objects on the ground.

In addition, because they were mechanical devices, barometric altimeters suffered significant lag, and they were not particularly accurate. Although Doolittle's specially made altimeter was reputedly capable of an accuracy of plus or minus ten feet, most pilots believed that plus or minus fifty feet was the best one could expect from service equipment.[7] A fifty-foot error was quite enough to cause accidents, either by causing a pilot to come in too low and land short of the field with the attendant risk of collision, or by approaching too high and overshooting the field, with the same result. Barometric altimeters, then, provided neither the kind of information pilots wanted, such as absolute altitude, nor sufficient precision to insure landing within the safe confines of an airfield. Pilots were unwilling to trust them when flying close to the ground.

The lack of absolute altitude indication in the 1929 experiments proved to be one of the most difficult problems to solve in the development of landing aids. Possible solutions using acoustic height finders (used by Germany's Zeppelins) and radio altimeters (demonstrated by E. F. W. Alexanderson) met with a lukewarm reception from the aviation community.[8] This disappointment was largely a reflection of the state of the art. These were bulky, heavy devices that required a dedicated operator. As aircraft of the period were small, and every pound consumed by operators and equipment was a pound of capacity unavailable for paying cargo, these devices also did not suit the interests of the air carriers. Alexanderson's device, for example, required the operator to transmit two signals in close proximity and then analyze the resulting graphical trace to determine the altitude. Then the operator could pass the information on to the pilot. The delay was enough to make the device too slow to assist with a landing. The Zeppelin acoustic height finder, although fine for airships that approached their mooring towers at only a few miles per hour, was also much too slow for aircraft. Hence, these ideas did not grab the attention of the community, and radio altimeters were developed for bombing use during World War II, while acoustic altimeters went with the Zeppelin into technological oblivion.

A different idea did catch the community's interest. Another scientist at the NBS radio section had conceived a means to provide another radio beam, angled up from the ground, which would essentially provide a safe inclined path to the landing field. Since the beam would be angled to clear all obstacles, an aircraft following it would also clear them, thereby eliminating the obstacle problem. Francis Dunmore's "landing beam," later called a "glide path," combined with Harry Diamond's localizer and marker beacons, was intended to provide a three-dimensional path in the air (see Figure 3.3). Two separate instruments were originally designed to display this path on the instrument panel, but after test pilots remarked on the difficulty of trying to coordinate the use of two special landing instruments and the Sperry artificial horizon, the bureau's researchers combined the landing instruments into one: the famous cross pointer (Figure 3.4). With the circle at the instrument's center representing the center of the two beams, and the intersection of the two pointers representing the plane's position with respect to the beams, the pilot simply had to keep the pointers crossed over the circle to effect a safe landing. The NBS first published the full description of its system in 1930 and tested it with the aid of the Aeronautics Branch at College Park, Maryland, and Newark, New Jersey, through 1934. The proposal drew wide acclaim from the aeronautical and engineering press. The German journal *Zeitscrift für das Weltflugwesen*, which published a lengthy abstract of the NBS publication, commented that this was the best solution to the blind landing problem yet devised, while the U.S. secretary of the navy wrote to the secretary of commerce to congratulate him for the "wonderful progress made" in blind landing after demonstrations during the summer of 1933. Various inventors continued to investigate other solutions, but none drew the attention that this idea did.[9]

Diamond and Dunmore intended the landing beam to work as follows. Figure 3.5 shows what the two engineers believed was the result of propagating a radio beam at 93.7 MHz from a set of horizontal dipole antennas arranged at a slight

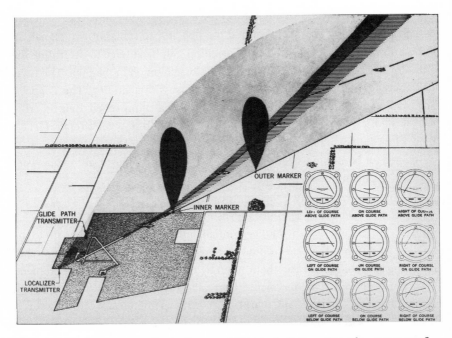

Figure 3.3. The beam pattern for the Bureau of Standards System. Note the curvature of the glide path. This diagram presents the approach path for the Indianapolis Experiment Station, and the image was probably made in 1936, when the Bureau of Air Commerce began to work with that system again. The sensing of the instrument displayed in the lower right is opposite that of the instrument in Fig. 3.4. The same image appeared in numerous publications between 1937 and 1941. National Air and Space Museum, Smithsonian Institution (SI 97-16983).

angle to the earth's surface. On the figure, the glide path is a surface of equal field intensity, leading to this type of glide path being called a *constant-intensity* glide path. Because the airplane's motion toward the transmitter increases the received field intensity, maintaining a constant received signal strength required the pilot to drop away continuously from the center of the beam. This descent would, according to the authors, inscribe a curve that was to touch the ground at a point about two thousand feet from the antenna.

Despite the widespread press attention, however, the NBS system remained an experimental apparatus. Test pilot Marshall Boggs's ability to land occasionally using the installation at College Park did not translate into an ability to do it regularly. He reported in a 1931 letter to the director of aeronautical development of the Aeronautics Branch, "I am beginning to doubt that the present approach

Figure 3.4. The cross-pointer indicator in an early form. The two zones at the bottom of the instrument face were colored green and red, left to right, respectively. These were later changed to yellow and blue because green and red could not be read under red night-time cockpit lighting. The sensing of this instrument is the inverse of that in Fig. 3.3. The aircraft is represented by the intersection of the two needles, with the center of the beam indicated by the large circle at the center. Author's collection.

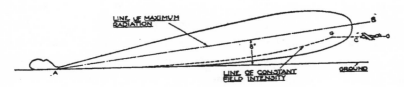

Figure 3.5. The constant-intensity glide path proposed in 1929. From Harry Diamond and Francis Dunmore, "A Radiobeacon and Receiving System for the Blind Landing of Aircraft," *Bureau of Standards Journal of Research RP 238* (October 1930): 924.

immediately adjacent to the field on the localizer and landing beam can be made consistently . . . It can be made occasionally, as I have already proved by my two 'blind landings' to date. It takes almost perfect conditions, however, to accomplish it,—i.e. no wind or a steady south wind, freedom from rough air or bumps, freedom from static and the perfect functioning of the radio transmitting and receiving apparatus."[10]

Further, he remarked upon the unsuitability of the College Park runway, which at 250 feet wide was too narrow. With a forty-three-foot-wide plane, only 103 feet to either side of the runway centerline remained, and he found that the aircraft (a Ford Trimotor, designated NS1) could drift through such a width before he could respond. He stated that at least a five-hundred-foot width was required to ensure reliable, safe landings. These conditions, and the desire of the Aeronautics Branch to test the system under commercial conditions, led it to arrange an installation at Newark during 1933, a field that had been built with four-hundred-foot-wide runways at the demand of Transcontinental Air Transport, which held the airmail contract for New York City. The airport was expanded during February 1934 to make "virtually the whole area" available for landings.[11] The increased area available made the landing problem much simpler.

Boggs's statement also points directly to one of the system's environmental problems. Bumpy air made flying the beams difficult but did not alter the beams themselves. Static conditions, the electricity buildup that occurs in storm conditions, on the other hand, could substantially alter the localizer course and the glide path angle, endangering the aircraft. The buildup of static electricity in the atmosphere "pulled" the localizer course toward the storm. An aircraft following the beam in would therefore not be aligned with the runway course, making a safe landing impossible.

In order to determine how well the system would perform at other locations and to begin training pilots in its use, the Aeronautics Branch arranged to install two copies of the system at Cleveland, Ohio, and Oakland, California, during 1933.[12] The two cities had the large, relatively unobstructed airfields that the system's long, low approach path required. Continued problems with the Newark installation along the lines Boggs had reported two years before, however, caused branch officials to reconsider that decision, and only the Oakland installation was made. The branch's officials were concerned that they would not have enough funds to improve the equipment if they made too many installations. The Oakland equipment had already been completed when that decision was made, however, and Oakland had agreed to pay the installation cost. The Aeronautics Branch could not back out of that deal, and Oakland got its equipment.

In January 1934, Eastern Air Transport sent pilot E. A. Cutrell to conduct further tests of the NBS's Newark experimental installation. Cutrell found that the localizer had a serious bend, probably caused by a nearby radio range, which, combined with the high speed of the Branch's Ford Trimotor airplane, made the beam difficult to fly. Cutrell accumulated nearly twenty-nine hours flying the system, but he did not make blind landings because he was unfamiliar with the plane's landing qualities. He also had "no first hand knowledge of how much abuse the airplane would stand."[13] He clearly expected blind landings using the system to be less than smooth.

The methodology used by Doolittle, Boggs, and Cutrell was essentially the same in all of these tests. They did not fly in bad weather. Doolittle and Boggs quite literally flew under a hood that was buttoned over the cockpit, while a safety pilot, who could see, monitored the approach. The safety pilot could take the controls in the event of a bad approach and fly the plane back to a safe altitude. This provided a significant psychological advantage that pilots flying under real conditions did not have. Pilots "under the hood" knew that their safety pilot would not let them crash. Without the reassurance provided by the safety pilot, pilots had to rely entirely on their instruments, which therefore had to represent a stable, reliable system.

The Bureau of Air Commerce's records, primarily because the agency dropped the NBS work in 1934 under pressure from the army and the airmail crisis, do not reveal the full magnitude of the system's problems. The U.S. Navy picked up the development work in 1933, however, via the Office of Naval Intelligence (ONI).[14] Seeking a landing aid for aircraft carriers and seaplane bases, ONI had a group called the Washington Institute of Technology, based at College Park, clone the NBS system. The Washington Institute was composed primarily of laid-off NBS researchers who had been collected by the director of naval intelligence and given a contract to complete and "navalize" the NBS system. The navy's records on it are the most extensive of all the agencies involved in investigating the NBS landing system, and they reflect the substantial technological difficulties the aeronautical community had with it.

The navy's clone was portable, with localizer and glide path transmitters in a trailer and the marker beacons in motorcycle sidecars. It was officially designated the YB system and was released in sealed containers in 1936 for commercial use under the name Air-Track system, named for its College Park manufacturer, the Air-Track Corporation. The Air-Track system has the distinction of providing the first blind landing to a passenger-carrying flight in Pittsburgh on January 28, 1938.[15] But the sealed nature of Air-Track, done to assuage navy national security demands, made the system impossible to maintain. It never became a serious competitor to other variations on the NBS's theme.

The navy's tests on the YB system began in 1935, with tests aboard the USS *Langley*. The tests were inconclusive but convinced the navy that although the system was not yet suitable for aircraft carriers it could be useful for seaplane bases. The navy continued developing the system and invited the chief signal officer of the army to attend a demonstration the following year. The report the two officers sent to the Signal Corps from the Aircraft Radio Laboratory at Wright Field provides an excellent review of problems encountered by the navy. They found that the glide path beam tended to vary its angle with the ground depending on the weather. On one particular day at the manufacturer's College Park testing facility, the pilots reported that as the ground dried in the sun after an early morning rain, the glide path angle slowly elevated so that by midafternoon, the path was unflyably steep. They also found that the localizer beam was pulled off course by the presence of static conditions near the airfield.[16] In other words, the system did not function reliably in bad weather—which was when it was most needed.

Nevertheless, the navy was satisfied enough with the system to establish an engineering group called Project Baker within the Patrol Utility Wing at San Diego in 1937. The navy's lack of concern with the above problems probably was due to the different landing requirements of seaplanes. Seaplanes did not have to reliably touch down at a well-defined point on the surface, as did land planes using runways. Course variations were therefore much less important. The San Diego location imposed an unexpected difficulty, however. The development team found that splits had developed in the localizer beam between the College Park tests and those done in San Diego the following year. Instead of producing a single navigable course along the approach path, the system was producing two courses. After months of fruitless modifications to the equipment, the group attached a graphic recorder to the receiver outputs in the test plane to find that the split course was caused by an interference pattern that surrounded the transmitter. Several more months of letter writing to various specialists eventually caused Project Baker's staff to decide that the interference was most likely caused by mountains more than one hundred miles away, in Mexico. They labeled the problem "uniformly recurrent multiple course phenomena," a suitably frightening-sounding label for a design problem experienced by all systems based upon low-frequency localizers. Because low frequencies could not be made directional, interference from any direction distorted the courses. The localizer splits delayed the navy's production decision into 1940, despite the advocacy of the commander of the Aircraft Scouting Force, Pacific Fleet, whose aircraft and pilots were being used for the tests throughout 1938.[17]

The navy finally accepted the equipment for its seaplane bases after receiving testimonials from several fleet pilots whom Project Baker had trained. Although the testimonials were written in support of the system, they also foreshadow one of the reasons for the navy's 1943 abandonment of it. Ensigns H. P. Gerdon and P. H. Craig thought that the long, low approach provided by the glide path beam's curvature was less safe than a straight one would be, due to the possibility of obstacles. Gerdon called the low approach a "mental hazard."[18] *Mental hazard* is an excellent term for what pilots experienced while using these variants of the NBS's system. Knowing that the beam led them at very low altitude over the surface, and knowing that the beam might not be stable at a given time, place, or weather condition, gave pilots good reason to be less than trusting. A minor change in the beam angle at such a low altitude could easily prove fatal to a crew. That danger imposed a mental hazard on pilots which caused them to mistrust the NBS system and its YB clone and drove radio engineers to look for ways of stabilizing the beams.

The instability problem doomed YB from the outset. The navy had not conducted site surveys of its intended installations because its leaders did not understand the new radio technology well enough to grasp the importance of site selection. Of its four highest priorities—Reykjavik, Iceland, and Sitka, Kodiak, and Dutch Harbor, Alaska—only Reykjavik possessed an approach path clear of nearby mountains. Mountains and mountainous islands in the approach paths of the Alaska facilities produced far more severe localizer distortions than had been present in San Diego, and which had not occurred at all in College Park. YB proved unusable at every Alaska facility where the navy tried to use it after the system's deployment began in 1941. By late 1943, the navy had abandoned the system completely.[19]

Both YB and its twin, the NBS system, suffered the same problem. Changing conditions on the earth's surface, from soil moisture to topographic features, caused changes in the glide path. Hills and even trees caused "bumps" in it that made flying the beam difficult, particularly for inexperienced pilots. Nearby mountains caused reflections that made the beams unflyable, while even distant ones could cause enough interference to be noticeable to pilots.

The instability of the localizer and glide path caused pilots to mistrust the system, and they would not use it as a blind landing system. Although I have focused on the system's instability, the major component of the NBS/YB system's failure was lack of trust. Landing an aircraft blind required an act of faith from pilots, and they would not place that faith, and their lives, in a system they knew was unstable. Getting pilots to trust the system meant finding a way to evade nature's inconsistency.

That engineers would eventually fix the instability problem was not foreordained, however, because the problem had nearly resulted in the NBS system's complete abandonment by the Bureau of Air Commerce in 1934. The poor results achieved by the system during the 1932 and 1933 trials had caused Eugene Vidal, Director of Air Commerce, to examine an army invention, hoping that it would be a more acceptable blind landing system. Capt. Albert F. Hegenberger at Wright Field had developed the army competitor between 1931 and 1933. Hegenberger was a veteran of the first nonstop flight from the West Coast to Hawaii and because of that feat was regarded as an expert in radio navigation.[20] In late 1933, Vidal asked the Air Corps to send Hegenberger and his system to Newark for testing by the Bureau of Air Commerce.

Hegenberger's system, formally named the A-1 system by the army but more commonly called the Hegenberger system, relied upon a radio compass to provide directional information to the pilot. Radio compasses, often called radio direction finders and now referred to as *automatic direction finders,* were a derivative of World War I radio research. Wartime demands for improved navigation had inspired radio engineers and radio physicists to investigate methods of locating radio transmissions, and one result was the deployment of a European network of radio direction finding stations that served as the primary means of navigation for aircraft.[21] These ground stations detected an aircraft's radio transmissions, triangulated with other stations, and then reported the plane's position to its pilot. What the U.S. Army Air Corps wanted, however, was a system that did the inverse. It wanted a radio station on the ground to activate an instrument in the airplane that the pilot could follow.

In 1931, the Air Corps hired Geoffrey Kreusi from Western Air Express. While at the airline, he had begun working on an airborne radio direction finder that the Air Corps leaders thought showed promise. It relied primarily upon a crossed-loop antenna on the aircraft's nose or belly to receive a radio transmission. Any radio transmission would suffice, so pilots could easily tune the device to commercial radio stations and use them as navigation aids if they wished.[22] The direction was accurate to within a few degrees—quite excellent by contemporary air navigation standards.

Kreusi's efforts at the Air Corps's Wright Field development center had succeeded by 1933, and Hegenberger decided to base a blind landing system on one.

He used two low-frequency, omnidirectional radio transmitters that he called *compass locators* to provide navigational references for the compass. Pilots were to fly between them using a specific pattern, shown in Figure 3.6. The two locator stations, which were truck-mounted, were to be placed just outside the airfield boundary (station A on the diagram) and 1.5 miles out (station B). Pilots set their radio receiver for station A's frequency and used the radio compass to fly toward it. Once the plane had passed over the inner station (indicated by the compass needle's sudden swing to point in the opposite direction), a pilot was to make a 180-degree turn, reset the receiver to station B's frequency, and fly to B. After continuing past B for about three more miles, pilots were to execute a standard rate turn and line up on B again, which they were to fly over at five hundred feet. The radio receiver was then retuned to A's frequency. Descending at four hundred feet per minute, pilots were supposed to fly over A at 150 feet and maintain the four hundred feet per minute descent until they touched down. The descent was measured with the Kollsman sensitive altimeter.

The Bureau of Air Commerce's testing of the Hegenberger system took place in early 1934, in the midst of what aviation historians call the airmail crisis. Eugene Vidal asked Foulois to send Hegenberger and his system to Newark to be tested alongside the NBS system installed there in February. In April, Eastern Air Transport's Cutrell flew the system, as he had the NBS system tests that January. He and his safety pilot completed thirty-seven landings in ninety-one attempts, a much better showing than they had achieved on the NBS system. On the strength of those results, and given the political situation, Vidal chose the army system. In June, he wrote to Foulois that the Bureau of Air Commerce wanted to adopt it as the civil standard. "Our investigation indicated clearly the superiority of the U.S. Army Air Corps system over the so called Bureau of Standards system for general adaptation to commercial aircraft," he wrote. His decision was based upon a report submitted by Major Snow, head of the blind landing program. Snow considered the tests conclusive and recommended that "any further tests of any blind landing system utilizing radio beams . . . be discontinued as wrong in principle and practice and of unwarranted expense in view of their past failure and slight future chance of success."[23] The army's Hegenberger, or A-1, system thus became the first officially sanctioned blind landing system in the United States.

Snow was certainly biased toward his own system, but his analysis was not wrong. The NBS system had not worked well in its official tests. He was correct that blind landings were impracticable using radio beams, and decades later that has not changed. Radio beams in that frequency range were and are not stable enough to permit safe blind landings. But the commercial pilots who were invited

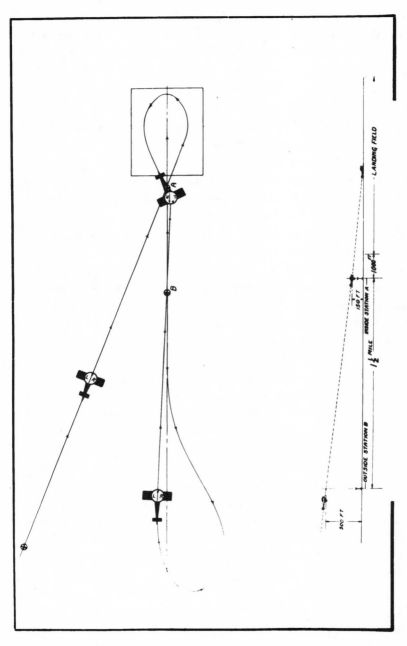

Figure 3.6. The Hegenberger, or A-1, approach procedure. This pattern took 1930s aircraft twenty minutes to fly, and because the aircraft was required to overfly the airfield during the approach, only one aircraft could be on approach at a time. This produced a very low landing rate. National Air and Space Museum, Smithsonian Institution (SI 98-15788).

to fly the A-1 system during 1934 believed that it was, in fact, inferior to the NBS system on which they had been practicing unofficially for the preceding year. They and their superiors were quick to take exception to Vidal's decision, and they inundated the bureau with letters explaining in great detail why they felt this way.

J. R. Cunningham, Superintendent of Communications of United Air Lines, wrote in October that his pilots did not "consider the sensitive altimeter dependable to the limits required for landing." In November, C. C. Shangraw of American Air Lines wrote to Major Snow that A-1 left a pilot "approximately 200' in the air over the inner guiding station and headed toward the field or runway . . . [having] no definite assurance of his height above ground, which obviously means that he has to feel his way down which required a longer runway or larger field than is normally available in commercial transport operation." Cunningham followed his October letter the following month with a detailed list of critiques that appear to summarize several of his pilots' opinions. The Kruesi compass permitted the plane to drift from the proper approach path after passing over beacon B, and in a blind landing the pilot had no way to detect that drift. And because the sensitive altimeter was typically off by twenty feet or so (and he reported deviations of up to one hundred feet in some aircraft), it was possible to land short of the field, or to land too far down field to stop before running out of space. Lack of a stable vertical indication, then, meant that the Hegenberger system did not define an adequate "point of contact" on the field. Several other pilots wrote the bureau with similar misgivings about the system's ability to provide safe blind landings at commercial fields.[24]

Clearly, the Air Corps and the commercial airlines had very different ideas about what worked as a landing system. That disagreement derives at least in part from the differences in their operating environments. The Air Corps in 1934 still relied overwhelmingly on all-over turf fields for its flying, for which an accurate point of contact was unnecessary. The corps also used smaller, lighter aircraft, mostly biplanes. The airlines had already junked such planes for faster, heavier all-metal monoplanes—Ford Trimotors, Boeing 247s, and DC-3s—and had demanded the adoption of surfaced runways to support all-weather operation of these relative monsters. One consequence of the airlines' adoption of surfaced runways is that landing on one required more precise guidance than the A-1 could give. What worked for the army's all-over fields thus did not work for the airlines and their runways.

The airlines were successful at getting Vidal's decision reversed, and the NBS abandoned the A-1 before installing it. Their efforts were helped by the Air Corps' excessive zeal to get the system into production. The corps delivered eighteen sets of ground equipment to the bureau, and all of them proved faulty. The bureau's officials vacillated for much of 1935 over whether to ask for money to fix them or to drop the whole idea but in the end decided to do what the airlines wanted. They let the program die. The Air Corps leaders saw that decision as a betrayal, complicating the relations between the two organizations right through World War II.

The airlines' reaction to the A-1 decision, in retrospect, was excessively negative. They were certainly correct that the A-1 was not an acceptable blind landing system, and during World War II the army air forces supported the development of several more systems in hopes of producing a more acceptable one. Yet the system's radio compass was an excellent navigational aid for finding an airport. The Bureau of Air Commerce's four-course ranges were effective en route navigation tools, but because their courses were not aligned with airport runways, they provided no help in making an approach and landing. Once off the range, pilots had to navigate to airfields by landmark and magnetic compass. Magnetic compasses were unreliable for that because of the aircraft's vertical motion during a turn. Landmarks, of course, were difficult to see through clouds, and in bad weather pilots had to risk flying close to the ground to find their way to a field. Some of the pilots who wrote to protest the A-1 system's adoption of a *landing* aid made a distinction that became very important later on: the radio compass was, they believed, an excellent *approach* aid because it made getting onto an approach course easy.[25] After World War II, the Air Line Pilots Association insisted that the compass locator stations be added to the NBS system's marker beacons for precisely this reason. Excessive devotion to blind landing in 1934 and 1935 thus drove the airlines and their pilots to reject a good *approach* aid because it was a poor *landing* aid.

The idea that Diamond and Dunmore had presented, then, a virtual path from sky to ground was so compelling that the commercial airlines and their pilots immediately adopted it as the best blind landing solution despite the minor problem that it did not work—yet. That it could be made to work, however, was an unquestioned belief. On that basis, the airlines managed to suppress the army's A-1 system, and United Air Lines took over the Oakland NBS installation and began working to fix the system's instability problems.

Recognizing that the NBS possessed most of the technical expertise on the system, United Air Lines' Cunningham approached the National Bureau of Standards and the Bureau of Air Commerce about obtaining their assistance in overcoming the system's problems in May 1935. He had made an arrangement with Eclipse Aviation, a subsidiary of Bendix Aviation, and he wanted to borrow some of the organizations' engineers to aid the two companies' development work. He agreed in advance to pay their usual salaries, making the deal no-cost to the government. The navy initially opposed the plan, with several officers concerned that the project would compromise the security of the YB system. Indeed, the secretary of the navy as early as 1933 had voiced national security concerns in his congratulatory letter to the secretary of commerce, and rightly so. The Soviet Union, Japan, France, Germany, and the United Kingdom all copied the NBS system, and Germany used the localizer as the basis for a "blind bombing" aid during World War II. But Adm. Ernest J. King, then chief of the Bureau of Aeronautics, convinced Adm. William H. Standley, the chief of naval operations, that not supporting the proposal would place the navy "in a position of encouraging the establishment of a monopoly" by making the Air-Track Corporation the sole manufacturer of a system that had, after all, originally been developed by a civilian agency for commercial purposes.[26] After nearly a year of delay, therefore, the alliance went ahead.

When United took over the NBS work in 1934, it immediately adopted a UHF localizer, primarily to reduce the system's cost. If localizer and glide path operated at the same frequency, only one transmitter and one receiver was necessary. That would reduce the ground installation's cost and the amount of weight that had to be carried by aircraft equipped to use the system. The choice of UHF for the localizer proved to be a good one, as we have already seen through the Lorenz case. It did nothing to help the already UHF glide path, however, and with its partner, Bendix Radio, United conducted an extensive investigation into the glide path stability problem. Primarily, the two companies focused on experimenting with antenna designs and with polarization. They found that a glide path using horizontal polarization was less affected by ground conditions than one relying upon vertical polarization, but simply changing the polarity was not enough to completely resolve the instability problem. They also abandoned the equal intensity glide path for an equi-signal type, in which the strengths of two differently modulated signals were compared. United Air Lines' engineers did this to help prevent changes in receiver sensitivity from changing the indicated glide path, thus eliminating another source of unreliability.[27] Yet neither they nor anyone else managed to completely eliminate glide path instability without shifting to microwave frequencies.

If the glide path could not be made completely stable, it could not be safely used for blind landings. But United's managers realized that truly blind conditions were very rare. They occurred less than 1 percent of the time, while ceilings below five hundred feet, the existing legal minimum, occurred between 5 and 10 percent of the time.[28] If one could produce a system that would permit safely cutting the minimum in half, well over half of all visibility-related delays could be eliminated. So United reduced its goals. It did not use the system for blind landings. Instead, it was satisfied to use the equipment as an approach aid, with landing to be done visually. United expected that the system would pay for itself in a few years by allowing completion of flights that would otherwise have to be canceled or diverted. Careful site surveys and equipment installation, combined with extensive pilot training, made this goal both reasonable and achievable.

Essentially, United redefined the problem to be solved as one of establishing an instrument approach system rather than a landing system. Without the demand for blind landing, one could reduce the twin needs of stability and precision. Precise electronic navigation was unnecessary if pilots were going to land visually. A system that could direct pilots to within a few degrees of the runway course was far better than nothing, and the remaining instability in the glide path was not severe enough to jeopardize aircraft that did not rely on it below two hundred feet. Safety would still be preserved, while most of the cost of weather delays would be avoided. This was, in fact, the ultimate "resolution" of the instability problem, but the Bureau of Air Commerce, the navy, and, to a lesser extent, the army required several more years of experience with failure to accept United's rather pragmatic attitude.

The United-Bendix system, therefore, was the most successful of the first generation of descendants of the NBS system, not because it was technically superior but because the organization using it recognized its limitations. It was installed at eight United fields in the United States, and between 1936 and 1937 more than three thousand landings were made with it "under the hood." It did not become more widespread because the equipment was expensive (around $10,000 per installation) and because there was no nationwide official standard. Airport managers waited for federal money to pay for these installations, and that, in turn, meant waiting for federal agencies to agree on which blind landing system to buy.

The attention given to the NBS research had not been exclusively American. While Japan and the Soviet Union hired U.S. engineers to duplicate the NBS system for them, in Germany the localizer–glide path–marker beacon idea, broadcast there by a translation of Diamond and Dunmore's 1930 article, had inspired radio engineer Ernst Kramar to develop his own version. The result, known as the Lorenz system after the eventual manufacturer, Carl Lorenz A. G. of Berlin, emerged from a joint project conducted by Lorenz, Telefunken, and the German aeronautical research laboratory DVL in 1932. The Lorenz system operated on 33.3 MHz, a frequency not available in the United States, where it was used by police and forest-fire patrols. It was markedly different than the NBS system in that it combined localizer and glide path into a single unit by alternately exciting two reflectors on either side of a vertical antenna with dots on one side, dashes on the other. This produced two elliptical patterns in space, with the major axes parallel to the on-course heading. The result was an equi-signal zone much like that produced by the NBS localizer. The glide path was in some ways a parasite: instead of producing one separately, a constant intensity line was simply selected from the existing "localizer" field pattern by the receiver. The United-Bendix team almost certainly got this idea from the Lorenz system. The system also employed two marker beacons operating on 38 MHz, with modulations of 700 Hz and 1,700 Hz to distinguish them.[29]

The company began testing the system in late 1932 at Templehof and began marketing it in 1935. The U.S. Army Signal Corps found that by 1936 installations were being made at Berlin, Frankfurt/Main, Hanover, Cologne, Hamburg, Konigsberg, Leipzig, Stuttgart, Danzig, Breslau, Munich, Nurnberg and Stettin in Germany, at Zurich, Switzerland, and Vienna, Austria, at Le Bourget in Paris, and at London's Heston aerodrome. Plans were being made for installations at Prague and Moscow, and one unit was shipped to Japan.[30]

Despite its rapid adoption in Europe, the Lorenz system shared the faults of its ancestor. In response to questioning by the U.S. assistant naval attaché for air in Paris, the director of Le Bourget aerodrome stated that "the curved beam did not maintain a definite course and shape due to meteorological conditions and building and hangar interference." The Lorenz company had such difficulty with its glide path that in a visit to the chief signal officer in February 1936, Carl Lorenz stated that his system did not include a glide path because it was still "experimental," despite his own sales literature and the testimony of Le Bourget's director.[31] As a result of the beam instability, pilots flew the Lorenz installations made during the mid-1930s with their glide path receivers disconnected, allowing them to use the localizer and marker beacons as a landing aid, but subject to the same lack of altitude indication that inspired commercial dislike of the A-1 system in the United States.

Even without the glide path, European pilots benefited from the Lorenz system, which simplified the basic task of finding the airport. It was especially useful given European air navigation practice. Instead of radio ranges for point-to-point navigation, European pilots relied on radio direction finding stations on the ground for position information. These stations triangulated on the plane's radio transmission, and then sent the resulting fix to the pilot, who then had to plot it on a map.[32] This was fine for en route navigation and in some ways was superior to the U.S. method, which provided no guidance for pilots flying outside the established airways, but it was much too slow for an approach. The procedure could put aircraft within the Lorenz system's beam, however, making the two an effective combination for making "instrument low approaches," although it was still not a safe blind landing system.

From the standpoint of the stability problem, the Lorenz system, in a sense, simultaneously represented a step forward and a step backward. Its UHF localizer was much less susceptible to atmospheric and long-range interference phenomena than the NBS system's low-frequency localizer had been. At the same time, it was more susceptible to local reflections (inelegantly called *multipath effects*). Local interferences, however, proved more manageable in the long run. Perhaps based on the success of the Lorenz system, UHF was quickly selected by U.S. radio engineers as the right choice for localizers because one could, by careful siting, prevent localizer reflections. Often that involved physical modification of airport surfaces, via movement of hangars, landscaping, and rerouting of traffic. But this was a far easier problem to solve than the "multiple recurrent course phenomena" the navy had run into in San Diego with its low-frequency localizer. That could only have been fixed by removing the offending mountains in Mexico, a far more substantial undertaking than merely moving a hangar or two. Hence, the UHF localizer required careful siting and perhaps some rearrangement of the local landscape, but nothing drastic. Protecting a UHF localizer from local interference was simply easier than protecting a low-frequency one from long-range interference.

By 1937, then, radio engineers and aviators in the United States and Europe had largely accepted that higher frequencies were better for landing aids. Environment-induced instabilities were more tractable with higher frequencies because they were local, and Chapter 4 examines engineers' attempts to move blind landing into the microwave realm. In that frequency range, they hoped to detach the signals from the earth's surface completely, making the environment irrelevant. Yet there was another way out of the stability trap, as the foregoing suggests: one could give up the notion of blind landings entirely. United already had, as had the European airlines. The United-Bendix, Lorenz, and A-1 systems were capable of producing reliable instrument approaches had U.S. authorities been willing to take a more pragmatic approach. The record shows, however, that neither the Army Air Corps nor the Bureau of Air Commerce were ready to give up. They still wanted blind landings.

CONCLUSION

The blind landing system pioneered by the National Bureau of Standards' Radio Section proved to have severe stability problems that led, in part, to its cancellation in favor of the army's A-1 system. Yet the mismatch between the capabilities of the A-1 system and the operating environment the airlines functioned within caused the airlines to resist the A-1. It simply was not and, as far as airline pilots could see, could not be made accurate enough to permit safe blind landings at most airports. The airlines' advocacy of the NBS system earned it a reprieve. The system's problems still had to be fixed, however, and that need drove organizations to seek more solutions.

That there was even a choice to be made descended from the NBS system's interaction with nature. If the original NBS system had been stable, there is little doubt that it would have been deployed beginning in 1933, and the army's challenger would not have been adopted. Nature thus posed a significant constraint on the evolution of blind landing systems. The need for some sort of system was a product of the collision between commercial aviation's demand for regularity and the earth's refusal to be consistent. That very inconsistency prevented the system and its UHF descendants from achieving the goal established for them by their inventors and institutional supporters: regular blind landings. That constraint, however, also motivated researchers to try new technologies and techniques in pursuit of the original goal. Nature's inconsistency constrained and drove the evolution of blind landing systems.

During the 1930s, therefore, engineers pursued several possible solutions to the beam instability problem, while policy makers sought to define an acceptable beam pattern. These two efforts affected one another. As the ability to generate straighter, more stable beams evolved, achieving agreement on a technical standard became increasingly difficult.

CHAPTER FOUR

The Promise of Microwaves

Commercial pilots' rejection of the Hegenberger system left a schism in the aviation community over how best to achieve blind landings. The Air Corps was satisfied with its radio compass approach, while commercial pilots wanted some version of the National Bureau of Standards' (NBS) tripartite system. In 1934, the community had not agreed on how a blind landing system should work.

Between 1934 and 1939, however, the Air Corps moved closer to the airlines' position, agreeing by 1939 that the three-part NBS model was, indeed, the best route to a solution. The remaining dissension was over details: how long should the glide path be? Must it be straight? What frequencies should it use? With consensus around a particular model of the solution achieved, the various organizations in the aviation community had common ground to negotiate over. Nevertheless, determining the system's details proved difficult.

The most vexing detail was the shape of the glide path beam. The Air Corps insisted on a straight glide slope, while airlines were satisfied with the "bent beam" idea and pushed for rapid deployment of such a system. The Air Corps, in turn, demanded a new system that relied on microwave radio, which was being designed in a unique collaboration between Civil Aeronautics Authority (CAA), the Massachusetts Institute of Technology (MIT), the Air Corps, Stanford University, and Sperry Gyroscope. The microwave system promised to provide the long, straight glide path that Air Corps officers believed was necessary, and they sought a way of delaying the airlines' drive for the bent-beam system. Gen. Henry "Hap" Arnold thus deployed the classic Washington delaying tactic—he requested an independent review of the various blind landing programs. In effect, this placed the decision in the hands of President Roosevelt, who turned it over to the National Academy of Sciences. Their recommendations would become policy until U.S. entry into World War II changed everything.

In the only existing history of this subject, William Leary has argued that Arnold's review was an "unnecessary complication" to the adoption process.[1] But Arnold had good reason, in the form of a new microwave-based system being developed for the Army Air Corps. As we have seen, the descendants of Diamond and Dunmore's system had serious stability problems, and Arnold believed the new microwave solution, not the troubled CAA system, was the correct choice. There is a tinge of bitterness in Leary's argument because Arnold's review had the effect of preventing the adoption of any commercial blind landing system until well after the war, costing the airlines money—and costing pilots' and passengers' lives. Arnold held a vision of a microwave future free of stability problems. The National Academy's review panel agreed with him.

DEFINING A "STANDARD" SYSTEM: THE RADIO TECHNICAL COMMITTEE FOR AERONAUTICS

In 1936, the Bureau of Air Commerce formed a committee to help it deal with problems it had found in the use of radio technologies in aviation. Called the Radio Technical Committee on Aeronautics (RTCA), the group was composed of senior people from the technical departments of aviation-related organizations. The Army Air Corps and Navy Bureau of Aeronautics each had two representatives, as did the Bureau of Air Commerce. The Federal Communications Commission and State Department each provided one member. From outside the federal government came representatives from the Air Transport Association, the trade organization representing the scheduled airlines, and the Air Line Pilots Association, the labor union representing commercial pilots. The chair of the committee was elected and tended to rotate among the group. Private, or general, aviation was not represented until after World War II.

One important note is the RTCA's legal basis of authority. It had none. The Bureau of Air Commerce had not sought, or received, blessing from Congress in the form of a law, or from the president in the shape of an executive order. The bureau's hope had been that by bringing together the full range of aviation interests from inside and outside government, whatever agreement was reached would have what Osmun has called the "authority of agreement."[2] The reasoning went that if everyone agreed to a standard, that standard would hold moral authority, even in the absence of legal authority. And absence of legal authority is precisely what existed: the Bureau of Air Commerce could recommend standards for airport equipment, but it had no enforcement power, and it was legally barred from funding airport improvements. It could not buy equipment to install at airports, blocking that possible route to establishing a de facto standard.

In addition to the RTCA's lack of legal authority, no other mechanism existed to coordinate aviation standards among Army, Navy, and Commerce Departments. The RTCA provided a forum for discussion of radio issues and for the negotiation of proposed standards, but for a standard to succeed, each member had to convince his (there were no women on the committee in the first twenty years of the RTCA's existence) parent organization to adopt the proposed standard. That proved difficult. Nevertheless, by providing a forum for discussion of needs and possibilities among various groups, the RTCA served an important role in the development of blind landing systems by establishing minimum criteria and, more importantly, by defining problems that still needed to be solved.

The Bureau of Air Commerce began to revisit the blind landing system question in 1936, after the airlines refused to use the Hegenberger system. It reinstalled the Newark equipment at Indianapolis and began once again to investigate the Bureau of Standards system as a possible solution to the blind landing problem. The rapid advancements made in the state of the art by other organizations, particularly the United-Bendix partnership, however, meant that the Newark installation was obsolete. This fact and the growing plethora of derivative systems led the bureau to ask the RTCA to examine all of the existing blind landing systems and, based on that, recommend a suitable standard. In this way, the bureau intended to capitalize on the knowledge gained and innovations made over the preceding few years by other organizations, without committing itself to any one company's design.

The RTCA's Subcommittee on Instrument Landing Devices, therefore, visited College Park, Maryland, to examine the Washington Institute of Technology's Air-Track system (the civil name for the navy's YB system), Kansas City, where United demonstrated the UAL-Bendix system, and finally Indianapolis, where on May 14 and 15, 1937, it tested the Bureau of Air Commerce's slightly modified system. More importantly, the Indianapolis visit also resulted in an extended testing of the Lorenz system, imported from Germany by International Telephone and Telegraph (IT&T), the parent company of Lorenz A. G. This first hands-on review of the only European instrument landing system brought extensive press attention, as well as detailed scrutiny by the Army Signal Corps.

The Signal Corps' representatives, Capt. George V. Holloman and Maj. F. S. Borum, argued that the curved, constant intensity flight path produced by the system required pilots to constantly change the aircraft's attitude and throttle setting during landing, which was in opposition to army training.[3] The army trained its pilots to land by maintaining the aircraft in a level attitude and slowly closing the throttles to maintain a powered glide into the airfield. This reduced the pilot's workload and was relatively easy to train pilots to do.

Pilot training was not the only problem the army foresaw in the curved glide path. The curvature placed aircraft at low altitudes (around fifty feet) at a half-mile from airport boundaries. The army believed that few of its airfields, and even fewer commercial airports, were sufficiently clear of obstacles to make this a safe approach altitude. The army's standard for obstacle clearance was to ensure a clear 20:1 slope for aircraft approaches, which was approximately a 2.8-degree glide. The Lorenz system provided a 1-degree glide from a half mile out, with the curvature increasing rapidly as distance from the airport increased. Beyond two miles out, pilots considered the path unflyably steep. They therefore had to intercept the glide path at roughly 700 feet at the outer marker 1.9 miles from the airfield, descend rapidly to fifty feet, and then level out into the final portion of the approach.

The two issues of pilot training and obstacle clearance mentioned in the two pilots' report was likely informed by a third, unstated, concern. Although the 16,000- to 25,000-pound aircraft commonly used by the air transport industry could negotiate the curved glide path in the hands of a skilled pilot, the army had much larger planes on its drawing boards. In fact, the 55,000-pound Boeing B-17 prototype had begun flying in 1936, and by 1937 Air Corps leaders were beginning to realize that aircraft of such size were not maneuverable enough to follow either this curved glide path, or the very short overall approach that the Lorenz—and to be fair, all other existing systems—provided. As a result, the director of the Aircraft Radio Laboratory, Col. John O. Mauborgne, added an endorsement to Holloman's report suggesting that the army needed to design, or have designed for it, a straight glide path.[4]

The two officers' criticisms were certainly valid. The training issue is difficult to assess, but the "army way" to land a plane became the "right way" after World War II, despite the army's complete replacement of its outdated instrument flying training program by a commercial one in 1941.[5] The army could have trained its pilots to use a curved glide path, but it rightly perceived no benefit in doing so.

The more significant issues were obstacle clearance and aircraft size. Obstacles in airport approaches were one of the pilots' union's biggest heartaches, and its president agitated in Congress to get that body to grant the Bureau of Air Commerce and its successor, the Civil Aeronautics Authority, the authority to regulate airport approaches. In 1937, in fact, pilots for Eastern Airlines and American Airlines refused to fly DC-2 and DC-3 aircraft into Washington-Hoover airport, citing the dangers posed by smokestacks, power lines, radio towers, and a highway that ran through the middle of the field. They were joined by Eleanor Roosevelt, who said she feared for her friends' lives in using the field, and a photo essay in the *Washington Times* showing the offending structures. The union also listed Fort Worth, Wichita, Chicago, San Francisco, Newark, Fresno, and Bakersfield as cities with seriously obstructed airfields. The Lorenz "bent beam" system would certainly not have been safe at any of these cities, until the municipalities agreed to remove the obstructions or move the airport.[6] The army's officers were correct in their belief that the system would not serve many airports without substantial modification of the airports' surroundings.

It is important to note that the Lorenz system was not the only curved glide path system. All systems that had been based on the Bureau of Standards work generated such a path and at roughly the same angle of inclination. Commercial operators had considered this desirable at first because following it reduced the descent rate of the aircraft as the plane approached touchdown, thereby reducing impact. When it had tested them in 1935 and 1936, the army had criticized this aspect of all these systems in its internal reports (which were apparently released to the airlines), but because the army was still clearly wedded to the Hegenberger system, which included no glide path at all, no one seems to have taken the army's criticism seriously. Colonel Mauborgne's suggestion that the army campaign for a straight glide path is the first indication that someone with significant authority in the army at least recognized the unsuitability of the army's own A-1 system. It also presaged a long-running argument between CAA, which wanted to deploy a bent-beam-type system, and the army, which demanded a straight glide path.

Following the demonstrations at Indianapolis, United Airlines hosted a meeting between representatives of the other major airlines, the FCC, the Bureau of Air Commerce, and Bendix Radio at its Chicago Office on June 23, 1937. At that meeting, a standard was agreed upon that was validated by the RTCA that September. Although it did not require a straight glide path, it stated that "study should be made of the possibility for obtaining a straight line constant rate of descent glide path." The airlines had also recognized the eventual need for a straight path, which might well have been based on their own expectations of much larger aircraft appearing in the near future. Paul Goldsborough, president of Aeronautical Radio, Inc. (ARINC), a company established by the airlines to make radio equipment especially for airline use, forwarded a copy of this proposed standard to the Air Corps headquarters. The proposed standard was published in the August 1937 newsletter of the Air Line Pilots Association, accompanied by a statement by R. T. Freng that "personally, I feel that this is the thing that we have been looking for a long time."[7] It was approved by the RTCA and published in the new Civil Aeronautics Authority's first technical report that October, along with a recommendation that CAA pursue development of the items listed under "Projected Developments," which included the straight-line glide path.

By late 1937, then, the Air Corps and commercial aviation groups had recognized that a straight glide path offered advantages to the curved one that all previous versions of the Bureau of Standards' system had generated. These advantages included a simpler landing procedure, greater length and consequently a longer approach from a higher altitude, and better obstacle clearance. But the ability to produce a straight path did not exist, and airlines believed that it was in their interests to adopt a curved path while a straight one was developed. It was over this point that the army disagreed, believing that waiting was better.

The response to the proposed standard within the Air Corps was mixed. The advantages of the proposed equipment were well understood, but criticism focused on the cost of the system. In a letter to the secretary of commerce, the secretary of war (very politely) pointed out that the shift to UHF frequencies that the standard imposed left Air Corps aircraft unable to use the system. Internal memos were much less polite, pointing out that the Air Corps, with more than 3,000 aircraft, faced a very expensive procurement program to outfit its planes in order to keep up with the Bureau of Air Commerce's improvements to the Federal Airways System. Not lost on the army men was the reality that the bureau was considering the needs of only about 300 aircraft operated by the commercial airlines, whose equipment the bureau did not have to pay for.[8] Hoping that in two or three years a straight path would be developed that might not be UHF, the Air Corps considered a temporary system unpalatable.

The major airlines were not only willing to pay for the receivers (which cost about $1,500 per aircraft) but were practically demanding the opportunity to do so. Despite believing that the equipment they wanted to buy in 1938 would be obsolete in at most four years due to the projected movement to a microwave system, the airlines were willing to install it. They wanted an instrument landing system to improve their profitability because delayed and canceled flights cost them a great deal of money. The airline demand for regularity that had led to surfaced runways also drove them to push for an instrument landing system. Airlines expected to make back the cost of receivers within a year of operations, as the additional income they expected to derive from use of the system offset the initial cost. Hence, the airlines' economic model indicated that rapid adoption of a "good enough" system was a better strategy than waiting for perfection.

The airlines therefore pushed the Bureau of Air Commerce to let a contract to someone to design a system that conformed to the published standard. In early 1938, the bureau awarded a contract to International Telephone Development Corporation (ITD), a subsidiary of International Telephone and Telegraph, to build a new system intended to meet the RTCA standard. The company's system was based upon two key patents held by one of its engineers. It consisted of a 110 MHz equi-signal localizer, a 93 MHz constant-intensity (curved) glide path, and two marker beacons. The installation at Indianapolis included two localizer transmitters and four glide path transmitters to allow approaches to be made from any of the four runway directions available at the field without having to move equipment. It employed a specialized array of five horizontal loop antennae for the localizer, which allowed variation of the course width by changing the amount of energy radiated by each of the loops. This invention was an important step in overcoming the problem of bent or split courses, because careful tuning of the system, combined with careful siting on the airfield surface, could provide a narrow enough beam pattern to avoid obstacles. The other significant patent was a transmitter bridge circuit that automatically compensated for aging of system components.[9] This allowed the system to maintain a constant output automatically, eliminating another source of system instability. Neither innovation dealt with the issue of glide-path curvature, however. As initially demonstrated in 1939, the Indianapolis system possessed the same two to three mile constant intensity curved glide path as its predecessor.

In September, the Radio Technical Commission for Aeronautics' Committee on Instrument Landing Devices met to formally evaluate the International Telephone Development Corporation's system at Indianapolis and the MIT system. Forty-six representatives of various organizations, including six from CAA, two from the Army Air Corps, and three from the Army Signal Corps met on September 13, 1939, at the CAA Experiment Station in Indianapolis. Six airlines were represented, as were MIT, Sperry Gyroscope, Bendix Radio, Bell Labs, Aeronautical Radio Inc., RCA, the International Telephone Development Corporation, the Washington Institute of Technology, and the Federal Communications Commission. Although there was no formally appointed member of the Air Line Pilots Association listed among the representatives, five commercial pilots were present, flew the system, and submitted an addendum to the report offering specific recommendations for modifications. E. A. Cutrell, who had participated in the

1934 tests of the A-1 and NBS systems, was one of these, as was the chief pilot for United Air Lines, R. T. Freng. The senior Air Corps representative was Maj. A. W. Marriner, of the Air Corps Communications Department in the Materials Division; the senior Signal Corps representative was Colonel Mitchell, director of the Aircraft Radio Laboratory. The organizer was Richard Gazely, head of the Technical Development Division of the CAA, and the chair was J. R. Cunningham, director of communications for United.[10]

The report of this committee indicates that Colonel Mitchell was active in supporting the army's interests, and in fact the chair requested his recommendation on the form in which the proposed standard was to be cast. Mitchell recommended that the standard be "left sufficiently broad to admit of many approaches to the problem, that is, that a performance specification only should be written."[11] The group unanimously agreed that a three-element system be chosen (marker beacons, glide path, and localizer, the elements of the Bureau of Standards' concept), and Mitchell suggested that the standard be written so as not to rule out use of a "heading device," by which he obviously meant a radio compass. The group again agreed and placed that device in the "projected developments" category as desirable, but not necessary.

The vital question of glide path shape, however, proved controversial. Preston Bassett of Sperry Gyroscope Company reported in an internal memo that "the meeting devolved into a wide variance of opinions, one section of which insisted on the adoption of the IT&T system for the next step, another group insisting that the system was impossible, and no blind landings could ever be made with it." Bassett also reported that there was a "general feeling among many of the pilots and operators that the bent beam was not the final solution."[12] Ultimately, the group chose to defer the decision to the pilots who were then out flying the system. The pilots' addendum to the report, accepted by the group as the committee's recommendation, specified that "a glide path intersection shall be obtained at 1500' altitude at a distance of 6 miles from the transmitter end of the runway . . . At the point of contact the glide path shall have an angle with the runway not less than 1 degree and not more than 2 degrees. The glide path shall pass through the following points—not less than 500' nor more than 700' altitude at a distance of 3 miles from far end of runway, and 1500' altitude at 6 miles."[13]

The commercial pilots had adopted an "almost straight" glide path of six miles length. This was well in excess of the two to three mile glide path that the Indianapolis system had achieved. Once again, the group placed the straight glide path in the "desirable developments" section of the standard.

Although the army members of the committee did not oppose the adoption of this standard, Mitchell was apparently not satisfied with it. He commented, in yet another attached addendum, that the Indianapolis system had merit as an experimental apparatus, especially since its ability to produce a straight glide path and a curved one would finally allow determination of which was safer. He thought that the straight glide path was too short and that the system should be extensively flown before final adoption. The army needed a ten to fifteen mile approach, instead of six miles. He obviously had in mind the Air Corps' new bombers, which were more than twice as heavy as the biggest commercial transport in service. He emphasized the need for a common CAA/army/navy standard and pointed out that the weight of the equipment made it unsuited to the army's smaller aircraft (which were, of course, the majority of army aircraft).[14]

It is important to note that the CAA does not seem to have intended the Indianapolis system to be the ultimate system. The organizer and head of the CAA's research division, Richard Gazely, pointed out, "The only really worthwhile opinion we believe is one based on extensive pilot experience. Only after 500 or more pilots have had a chance to observe the behaviour of the equipment under every condition encountered in routine operations and only after these pilots tell us that they would trust the lives of their passengers and their companies' equipment to the guidance which a system gives them will we permit ourselves to believe that the system is good. But to get this kind of pilot opinion requires something more than a laboratory installation. A rather costly large-scale installation is necessary."[15]

He concluded with the hope that the experience gained would allow the system to be "improved out of existence."[16] The CAA intended to buy ten of these systems in 1940 so as to enable this extended testing to take place in a variety of locations. It also planned an additional purchase of fifteen for 1941.

The CAA was able to commit to a standard and buy equipment for airports in support of it because Congress had given it new legal authority in August 1938 in a major reorganization.[17] The CAA's legal authority to adopt a standard for civil aviation did not supplant military prerogatives, however, and it had to negotiate with the War and Navy Departments to achieve military/civil standardization. Widespread belief that a single standard for military and civil aviation was necessary, combined with a technical desire to eliminate the ground-induced stability problem, led it and the RTCA to view the Indianapolis system as transitional. Any adoption of it was to be temporary, to produce the sort of information regarding performance at many locations which all previous testing had not been able to provide, while continuing to investigate methods of producing a straight glide path. Following the RTCA's recommendation, the CAA continued its re-

search program to develop the straight, microwave path that was clearly the solution preferred by many of its members. The RTCA's standard, then, was a
means of establishing a coherent development program while providing limited
service to the airlines.

THE MICROWAVE FUTURE

The root of the glide path stability problem was that Diamond and Dunmore's
"landing beam" was not a beam at all. Instead, the glide path antenna's radiation
pattern relied upon the earth's surface to act as a reflector. That meant surface
conditions affected the radiation pattern. Changes in the soil moisture content of
the surface altered the soil's conductivity and thus the glide path's transmission
pattern. That, in turn, changed the glide path angle. This was why the army's 1936
tests found that the glide path angle changed as the ground dried out after a rain.
It also meant that changes in surface conductivity caused by large pipes, underground electrical conduits, and similar artificial structures caused "bumps" in the
glide path. Fixing the glide path, then, meant finding a way to detach the glide
path's radio propagation from the earth's surface.

Wilmer L. Barrow, an MIT engineer, had discovered in 1938 that radio waves
could be propagated directionally, without reflection by the ground, by using
metal horns. In other words, the use of these "horn radiators," as Barrow called
them, could generate radio beams independent of the earth's surface. This was
clearly a potential solution to the instability problem, but it brought with it a substantial difficulty: the horns were very large. The size depended upon frequency
in an inverse relationship.[18] Higher frequencies needed smaller horns.

In 1936, CAA engineer Irving Metcalf had proposed a blind landing system
based on three beams of energy, which would appear as three spots on a cathode
ray tube display. Initially, he had believed that infrared energy would be the best
choice, but after he consulted with Edward Bowles at MIT he decided microwaves
were a better choice. Working at the Round Hill experiment station, Bowles's research group had demonstrated that infrared would not penetrate all types of fog
and mist.[19] Microwave radio seemed a better solution.

Metcalf was able to convince his superiors to fund a twelve-month contract
with MIT to build a prototype blind landing system based upon Barrow's horns
and microwaves at fifty-centimeter wavelength.[20] The horns were still large but
much more manageable than they would have been at lower frequencies. Ultimately, the Round Hill group devised a system that produced four beams instead
of three. By overlapping two vertical fan-shaped beams along their flat faces, the

system established a plane that provided an indication of proper course. Similarly, it overlapped two horizontally fanned beams to produce a plane for vertical navigation reference. Although all four beams were transmitted at fifty centimeters, each was modulated with a separate audio frequency, which was then separated by filters in the receiver.

The resulting system had two great advantages over the Bureau of Standards system's UHF descendants: the beams were perfectly straight, and they were nearly immune from changes in ground conditions. It therefore answered pilots' demands for a more stable system. A successful demonstration of the system in early 1938 led to a six-month extension of the bureau's contract, so that Bowles could finish the report he was required to submit. The system had one major drawback, however: the cluster of triodes which generated the microwave signal could only produce about three watts of power, resulting in a useful range of only about a mile. More power was clearly necessary for a useful blind landing system. Fortunately, a potential solution already existed, a new vacuum tube called a klystron.

Physicist William Hansen and two other researchers, Russell and Sigurd Varian, developed the klystron in the Stanford University physics department. Stanford's physics program was oriented toward research into high-powered devices. In 1936, Hansen had developed a microwave tube called a rhumbatron, but this did not produce much power. Department chair William Webster had unsuccessfully sought money for a "million-volt X-ray" device in 1937, and after this failure Hansen turned toward research into techniques that might be less expensive. Russell Varian, who had been a student of Hansen's before going into private enterprise, was also interested in this research area, and he kept in touch with Hansen. According to Russell, Hansen wrote to him in 1937 that hollow resonators seemed to be very efficient at generating high-frequency radio energy.[21] This was the basic idea behind the klystron.

Russell and his brother Sigurd, a Pan Am pilot, had also corresponded about possible uses of microwave energy, including aircraft detection and blind landing. Sigurd had expressed concern over the concentration of power—particularly air power—in hands of European dictators. The Spanish Civil War had made clear the future importance of air power, and Sigurd believed that aircraft could easily bomb targets via "blind flying" techniques, while the targets could not defend themselves against the invisible aircraft. Microwave-based radio locators might resolve this problem. In September 1937, the two brothers went to Stanford and argued for a project to develop a resonator-based tube that might produce the high-energy microwave radiation an aircraft detector would require.

Sigurd was particularly adamant about it, according to Russell, and they succeeded in persuading Hansen. Hansen got Stanford president Ray Lyman Wilbur to support the effort with a $100 grant. The university's board of directors approved the arrangements in early October.[22]

The Varians quickly succeeded at assembling an apparatus that produced microwave energy and, nearing the end of their finances, tried to produce interest in the device by visiting the local navy and CAA offices. Irving Metcalf happened to be in town, and he and Hugh Willis of Sperry Gyroscope were immediately interested in seeing it in action. Metcalf realized that combining the klystron with Bowles's construct at MIT was the obvious route to making a stable, useable glide path. Willis perceived a new business opportunity, and he quickly arranged a deal with the Varians and Stanford to take over financial support of the klystron effort.[23]

The microwave work at MIT and Stanford had also attracted the attention of the U.S. Army Signal Corps. Col. John Mauborgne, who had been director of the Aircraft Radio Laboratory at Wright Field, had been promoted to brigadier general and assigned chief signal officer of the army in early 1938. In April of 1938 General Mauborgne visited MIT, where he met with Karl Compton and Edward Bowles and witnessed a demonstration of the CAA-MIT microwave landing system.[24] He came away from that meeting convinced that it was the solution to the army's need for a straight glide path, and he arranged to send people from the Signal Corp's Radio Laboratory to MIT to learn more about it. Bowles's assistance, in turn, led the Signal Corps to establish an in-house research program into microwave landing systems at Wright Field. At first, the army was only interested in the glide path part of the system and arranged to borrow a localizer from the CAA to use in their tests.

Mauborgne's replacement as director of the Aircraft Radio Lab, Lt. Col. Hugh Mitchell, kept him informed of new research through correspondence with one of his staff members, Col. Luis Bender. Mitchell heard of Stanford's klystron from Metcalf and had corresponded with Webster about it. Mitchell had then described the device to Bender in a handwritten note, emphasizing the importance of this device for blind landing of aircraft, as he knew that his boss was interested in the subject.[25] He believed it was the best possibility for extending the range of Bowles's glide path from a mile to the minimum of ten that the army wanted.

Mauborgne arranged to borrow one of the prototype klystrons from Stanford for use in the army's copy of Bowles's glide path transmitter, believing that joining the two would provide a useable range. The army first flew the klystron to MIT, however, where Bowles was able to use it for a few weeks before the army's equipment was ready. Initial experiments proved very exciting, but flight tests originally scheduled for May 1939 had to be postponed into June because the klystron had to be sent back to Stanford for repairs. The research projects at MIT and Wright Field suffered from frequent problems with the prototype klystrons, forcing them to send the devices back and forth between MIT, Wright Field, and Stanford, which considerably delayed their work.

The shortage of klystrons in early 1939 led the CAA, MIT, and the army's Aircraft Radio Lab to cooperate at MIT. MIT, which expected to get a new klystron before the army's was repaired, approved an army request to fly MIT's system once the klystron arrived. The CAA had renewed its research contract with MIT, although its efforts in the microwave area were waning as the UHF system it had contracted for with ITD neared completion at Indianapolis. The Sperry Gyroscope Company, which had established a development laboratory for the klystron in San Carlos, California, was also informally involved in the undertaking at this point.[26]

The cooperation was shaky, however. General Arnold had written to Clinton Hester, administrator of Civil Aeronautics, proposing a cooperative development effort in the microwave field. Hester happily agreed to participate and even to provide the equipment.[27] Arnold then demanded establishment of an overall policy of cooperation before agreeing to a cooperative program in this specific case, however, which Hester appears to have been unwilling to commit to. Relations between the two organizations deteriorated, while Bowles's group continued its work at MIT and the Signal Corps prepared its own version of MIT's system at Wright Field. But the research at MIT and Stanford had produced a new set of technologies that would satisfy the army's requirements for a straight glide path. The remaining issue for the army was how to get everyone else to adopt what it wanted.

THE NATIONAL ACADEMY OF SCIENCES:
VALIDATING A RESEARCH PROGRAM

The completion and relatively successful testing of the International Telephone Development Corporation's design at Indianapolis during the summer of 1939, while the army's enthusiasm for microwaves was still growing, set the stage for what other historians have presented as an army attempt to block the RTCA's standard. Wilson and Leary have argued, based upon CAA records, that General Arnold sought to prevent the CAA from carrying out the RTCA's plan by calling for a National Academy of Sciences (NAS) review, apparently in the hope that

NAS would recommend something else. Army records suggest, however, that Arnold's memos were misunderstood in the White House and that Arnold had no intention to block limited commercial deployment, which was, after all, the RTCA's own plan. Instead, Arnold sought validation of his belief that microwaves were the best solution to the glide path problem and therefore of the Signal Corps research program. Once he had that validation, he intended to insure that the CAA conformed to the limited deployment of the RTCA's system, as he believed it had agreed to do.

Arnold, after a conversation with the assistant secretary of war for air in August 1939, had written for the secretary of war's signature a memo to President Roosevelt. In it, Arnold requested that the NAS be asked to evaluate the existing blind landing systems. They were, he argued, a "disinterested group of distinguished scientists who would serve much as a court of justice serves in our system of jurisprudence." The Air Corps had gotten good advice from them in the past, Arnold believed, and he thought that their national prestige would end all "quibbling or contraversy [sic]" once they had announced their decision. Arnold had begun working with members of the National Research Council in 1936, as Gen. Oscar Westover's assistant. That experience had convinced him that scientists were willing and able to advise the Air Corps on new technologies in which it had no expertise.[28] The NAS was therefore a reasonable place for him to turn to resolve what he perceived as a divergence between the needs of the Air Corps and the plans of the CAA.

Arnold's letter to President Roosevelt got an immediate response. In a letter dated the same day, August 30, Roosevelt asked Paul Brockett, the executive secretary of the National Academy of Sciences, to look into the problem of standardizing upon a common system. Brockett turned the letter over to Frank Jewett, then president of the NAS, who responded on September 1. Jewett agreed to appoint a committee to look into the problem, with the caveat that it might not be possible to attempt complete standardization.[29] Jewett's initial sense, that standardization might not have been possible given the state of the art, is certainly borne out by the RTCA's construction of an interim standard.

Jewett chose Vannevar Bush to chair that committee and assembled a number of prominent researchers to serve on it. The conference committee consisted of Oliver Buckley, executive vice president of Bell Labs; electrical engineer Dano Gunn; mechanical engineer W. F. Durand from Stanford; physiologist Joseph Erlanger of Washington University; telephone engineer Bancroft Gherardi, who was retired from AT&T; biological chemist Lawrence Joseph Henderson of Harvard; aeronautical engineer Jerome Hunsaker at MIT; and physicist Max Mason of the California Institute of Technology.[30] Bush also wrote to the various aviation interests, including the CAA, the Air Corps, the navy, the Air Line Pilots Association, and the Air Transport Association, seeking their positions on the issue.

In October, Charles Stanton, acting administrator for Civil Aeronautics, responded to Bush's request with a long explanation of the CAA's current policy. Stanton explained that the CAA intended to install ten sets of equipment at important airports, while continuing to improve the equipment. The CAA recognized the need to familiarize pilots with instrument approach procedures well in advance of attempting blind landings in routine operations. He asserted that the CAA intended to continue funding MIT's research into a microwave-based system. A reliable, manufacturable microwave system, he believed, was three years away, and because airlines expected aircraft radio equipment to become obsolete every three to four years, he felt no concern that deploying a UHF system would prevent adoption of a microwave one once one was available. Air carriers, he pointed out, had "shown eagerness to accept the use of new equipment which tend to increase the regularity with safety of scheduled operations."[31] Stanton argued for the approval of ten ground stations.

Bush received similar advice from Edgar S. Gorrell, president of the Air Transport Association. Gorrell had written to each of the major airlines to get their support for Bush's investigation, and expressed their "sincerest cooperation with the National Academy of Sciences in the subject of one instrument landing system for common adoption." Gorrell also pointed out that the CAA had a project under way to produce a standard landing system through the RTCA and attached a copy of the September 14, 1939, proposed standard to his letter for Bush's edification. The air carriers, he said, were convinced by their many years experience in testing various systems that the CAA's new system was the best that they had seen, and he urged that "nothing be done to retard the program."[32] The airlines, he said, were unanimous in their wish that the CAA proceed with the initial ten installations. The airlines were clearly satisfied that even as an interim system, the RTCA's proposal would provide sufficient economic benefit to justify their costs.

Bush's committee had no significant funding, and it met only once, on October 14 and 15, 1939. The group saw presentations by the army, the CAA, and the navy. It drew up a draft report on the fifteenth, and the committee decided that additional information should be gathered from the various involved agencies. It also decided that Bush should personally fly each of the systems in question, particularly the CAA's Indianapolis installation and the MIT microwave experiment. He was to keep the committee informed by letter.

At the meeting, the committee focused primarily on CAA's system and MIT's microwave project. The report it drafted on the fifteenth proposed continuing the CAA's plans to install ten copies of the Indianapolis system, with the caveat that it be modified to meet the six-mile, 1,500-foot approach that the pilots had specified the month before but that ITD's engineers had not yet managed to achieve. The draft also supported the army's desire for a straight path and advocated continuing microwave research as the best means to produce it. In short, the NAS's draft report echoed the RTCA program almost exactly, with somewhat more emphasis placed on microwave development than in the RTCA document.[33]

As the largest unresolved issue was the straight versus curved glide path problem, Bush sought more information from various parties. He received a letter from Charles Stanton on November 3 notifying him that the Indianapolis glide path was now in conformance with the RTCA specification. Stanton also forwarded a letter from Colonel Mitchell to Richard Gazely, which stated that the army found the CAA's specification acceptable for test purposes, but for army use a completely straight path, inclined at 3 degrees, was necessary. The colonel reiterated that he supported the installation of ten sets of the CAA's system for pilot training and familiarization, but deployment should not go further until the system's real world performance was well understood.[34]

Bush also received a long letter from Bowles that discussed the Indianapolis system and the MIT microwave experiments. Bowles told Bush he had spoken with William Jackson from CAA's Technical Development Section and had been told that the straight six-mile glide path that had been produced had not yet been thoroughly tested for straightness and "freedom from irregularities." Further, Bowles had been left with the impression that the straight portion could not be substantially lengthened with the existing equipment because this was done by moving the transmitter equipment further to one side of the runway. The further away from the runway the equipment was moved, the greater the probability of interference with the radiation pattern from reflections and other surface conditions. Although Bowles did not discuss the issue in any detail, he stated that ITD's engineers believed that an equal signal glide path was possible at UHF wavelengths.[35]

Bowles argued explicitly for an equal signal glide path in his much more detailed discussion of the microwave system. "I believe that the groups interested in the instrument landing of airplanes subscribe to the idea that the ultimate system will be a microwave system and that the value of a longer wave system lies principally in the fact that these systems are now in such form that at least they offer immediate means for the training of commercial pilots in the technique of instrument approach and instrument landing."[36]

He added that the constant-intensity glide path at Indianapolis "must sooner or later degenerate into a curved path" if the CAA made further attempts to lengthen it. He had demonstrated already that microwave radiators could produce a straight equal signal glide path by means of a sharp ultra-high-frequency beam.[37] The army, he thought, agreed with his point of view and were working on such a system at Wright Field. Finally, he asserted that an equal signal glide path, whether UHF or microwave, was most likely to resolve the straightness problem and the instability problem because it was an "actual path in space" that could be made independent of transmitter output and receiver sensitivity.

In a prescient paragraph, he also argued that the "flare" at the bottom, which had been specified by the pilots to reduce impact upon landing, would be unnecessary and should be dropped because the system would only be used as an *approach* system, not a landing system.[38] If the system was only flown as an approach system, with the pilot making a visual landing, any desired flare out could be done visually. Bowles's estimation of the unfeasibility of a true "blind" landing system was no different than that of United Air Lines, which had already recognized that blind landings belonged to the remote future, if they were possible at all. Within a year, the army also picked up on the improbability of truly blind landings and tried to change the name given to this class of technologies from "instrument landing systems" to "instrument approach systems."

Bowles argued, finally, that the fastest path to a microwave system was a UHF/microwave hybrid, utilizing the UHF localizer of the Indianapolis system and a microwave glide path, put together by a commercial company. He pointed out that RCA, General Electric, and Sperry Gyroscope were working on equipment suitable for this task. He also contended that the "ultimate" system should be all microwave, to resolve the remaining problems with bends in UHF localizers, and should employ a ten centimeter wavelength, instead of the fifty centimeters that he and the Aircraft Radio Laboratory had been working on. He believed that this could reasonably be done in two years.[39] The Aircraft Radio Lab had already taken this approach at Wright Field, probably on Bowles's recommendation.

Having digested all of the arguments presented by his interested parties, Bush flew off to try out the various systems. At the Aircraft Radio Lab, he was able to try a Bendix system, the old Army Hegenberger system, and at nearby Patterson Field, the radio lab's prototype microwave system. He wrote the committee on

November 20 that when flying the Bendix system, he was "further impressed with the desirability of a long, substantially straight path, although the system in general operated successfully and satisfactorily." On the old army A-1 system, his pilot, whom he called "one of the most skillful and experienced," made two approaches and missed both. Bush was "completely convinced" that this system made "too severe demands upon the pilot," and this was reason enough for its abandonment. At Patterson Field, he found the microwave equipment to be "in experimental condition only." He was unable to fly the system because both of the available microwave receivers broke during his visit. He was convinced of the microwave system's "great future promise" but was clearly less optimistic than Bowles had been about its nearness to commercial utility.[40]

At Indianapolis, Bush found that William Jackson's group had done an excellent job on the equipment. His pilot made satisfactory instrument landings from all four directions. He judged it "completely satisfactory for commercial use." He agreed that the substantially straight part of the glide path was six miles long and admitted that while the army wanted an even longer glide path, he was "inclined to believe that much more is not really necessary."[41]

The issue of the length of the glide path is difficult to evaluate. The length to be chosen depended on how air traffic was managed and what sort of aircraft were using a facility. In commercial operations, especially during poor weather, aircraft were typically stacked above a radio marker beacon at some distance from the airport, with the most recent arrival at the top of the stack, and the next aircraft to be cleared to land at the bottom. Since commercial aircraft flew at a maximum of 10,000 feet (they were unpressurized), and aircraft were separated by at least 1,000 feet within a stack, having the base of the stack at 1,500 feet allowed a nine-layer stack. The width of a stack, and therefore its overall volume, was defined by aircraft performance. Faster aircraft, and heavier aircraft with poorer turning abilities, meant a bigger stack that then had to be further away from an airfield in order to prevent congestion close to the field. Because Bush was not a pilot and flew these systems in commercial aircraft with pilots familiar with commercial needs and in good weather, the six-mile glide path seemed perfectly reasonable to him. It meshed with the current suite of technologies in use by the airlines.

By 1939, however, the Air Corps was getting aircraft that flew higher, weighed more, and were faster than anything in commercial use and therefore demanded larger, more distant stacks. That in turn meant a longer approach. The need to stack those planes was reflected in the Air Corps' demand for a longer glide path. It is true that the Air Corps could probably have lived with a localizer-only approach for the first few miles of a descent (and, in fact, it did not have a glide path until 1944), but that largely defeated its purpose in adopting a glide path, which was to reduce accidents through improved vertical guidance.

The army wanted its pilots to pick up the glide path fifteen miles out and fly the twin beams in, instead of picking up one beam at fifteen miles and the other at six miles. Since the cross-pointer instrument displayed vertical and horizontal position, it provided a vertical indication even if it was not receiving a glide path signal—the elevation needle did not disappear in the absence of a signal. Instead, no signal resulted in a "fly down" indication, which a pilot would have had to ignore until approximately six miles out, with no certain indication of when he should start receiving the signal and therefore start paying attention to the elevation needle. Hence, the Air Corps' demand was driven by the performance characteristics it had designed into its new bombers, combined with the need to provide its pilots with a positive instrument indication. Bush could not have realized this from his own experience with these systems, but he deferred to the army's demands in his final report.

After returning to Washington, Bush made some minor changes to the draft report, the most significant of which were a specification that the army and the CAA needed to agree on a common sensing of the cross-pointer instrument through the RTCA, and a statement that the committee did not feel that a proper solution lay in the low-frequency direction. He requested that the committee telegraph their approvals by November 24 so that the report could be submitted promptly.

Bush apparently received the approval quickly, for the report he sent to Jewett bears the date November 21. Jewett, in turn, transmitted it directly to President Roosevelt. Roosevelt had his personal secretary, Gen. Edwin Watson, send copies of it to the army, the navy, and the CAA on December 5. The CAA responded with a highly laudatory letter to Bush, thanking the committee for its efforts in investigation of this "highly important and complex subject."[42]

Although the report was eventually approved unchanged, the process of approving it took longer than drafting the report itself. The first stumbling block was the Civil Aeronautics Authority, which believed it had gotten exactly what it had wanted and pushed the National Academy of Sciences to make the report public immediately. The academy's executive secretary wisely checked with the White House and found that Roosevelt considered it a confidential report until he, personally, chose to release it. The CAA had apparently not been quite so circumspect, and details found their way into the press anyway. In Brockett's words, this had left the White House "just a little annoyed" with the CAA.[43]

The secretary of war, Harry Woodring, responded on December 28 to General

Watson's memo. The War Department, he said, concurred with the NAS report, and he emphasized that the department considered the plan for future development it proposed "to be of material assistance in guiding the development towards standardization of an instrument landing system available to all aviation services."[44] The records show that Arnold himself drafted Woodring's reply. Arnold, therefore, was primarily interested in the report for its validation of the army's research plans, which were clearly present in the report's call for microwaves.

On the day Arnold wrote Woodring's response to the NAS report, he also acted upon a memo initiated by the chief signal officer in October. General Mauborgne had requested Arnold's opinion on whether to pursue the army's microwave research by letting a contract to Sperry Gyroscope Company for a complete microwave system, based upon the klystron. Arnold had clearly held onto the memo while waiting for the NAS report. In his answer to Mauborgne, Arnold requested that the Signal Corps "expedite" its research program and then quoted the NAS report's microwave section verbatim.[45] His holding of the chief signal officer's request and his reliance on the Bush committee's findings suggest that Arnold had already come to rely heavily on scientists' recommendations in subjects outside his expertise. He sought validation of the RTCA's program through the NAS and wound up causing a great deal more trouble than he intended.

The secretary of war's concurrence was not enough to establish the academy's report as a formal policy. That there was also no policy to guide its adoption led to a great deal of confusion. Because War and Navy were Cabinet departments and the CAA was an independent agency, no authority existed to approve and enforce the report's recommendations other than Roosevelt himself. The White House did not immediately recognize this, and after Roosevelt received the secretary of war's statement of approval, he asked his secretary, "What do we do next?" Watson was equally in the dark, and he wrote to the secretary of war asking his opinion, who then told the chiefs of the Air Corps and Signal Corps to submit their ideas on how to proceed.[46]

With much of official Washington still on the traditional Christmas hiatus, neither the chief signal officer nor the chief of the Air Corps responded in person. Instead, their executive officers responded, most likely after consulting with their bosses. In any case, the executive officer of the Signal Corps, Col. Clyde Eastman, suggested that the president approve the report and "furnish copies of it to the interested agencies for their guidance." The executive officer of the Air Corps, Maj. C. E. Duncan, concurred. The secretary of war accepted their recommendation and so recommended to General Watson.[47]

By putting the decision back in the White House, the Air and Signal Corps placed a technical decision in the hands of nontechnical people. Roosevelt relied heavily upon Watson's advice in army matters, and Watson, a retired infantry officer, did not understand the issues. A series of letters between Watson, Woodring, and Arnold ensued as Arnold attempted to explain to the others what the Air Corps did and did not want. A particular set of personal memos between Arnold and Watson had convinced the confused Watson that Arnold disliked the CAA's plans as embodied in the NAS report, and that perception was transmitted to Clinton Hester at the CAA. The CAA's records therefore document army resistance that did not exist.

The confusion arose from a combination of Watson's lack of comprehension of the basic issues and an imprecise memo from Arnold. Watson had asked Arnold for his opinion of the NAS's report in a memo dated February 6. Arnold had responded:

Dear "Pa:"
The Air Corps has no fault to find with Dr. Bush's proposed solution. We still, of course, believe that the system which we developed is better suited to meet Air Corps needs, but we realize fully that it is an impossible situation to have different aeronautical agencies each develop their own systems, no two of which any one airplane could use. It is absolutely essential that one common system be in use by the whole aviation industry and we are perfectly willing to give and take and compromise in order to arrive at that universal system.[48]

Watson seems to have understood this memo as supporting the army's old Hegenberger system over the academy's recommendations. That interpretation would certainly have indicated hostility on Arnold's part towards the report's recommendations, as Bush and his committee had clearly consigned the Hegenberger system to the scrap heap.

The context suggests that Arnold had meant to support the army's microwave system with this memo. Arnold was well aware of the Aircraft Radio Lab's work with Bowles on that prototype. He had sent Air Corps pilots to MIT to test the microwave system that the CAA had funded there. He had been satisfied enough with the Aircraft Radio Laboratory's progress with the system to ask the chief signal officer to expedite a procurement contract with Sperry Gyroscope for a manufacturing prototype less than two months before. Finally, he had personally signed the Air Corps' approval of the NAS report and drafted a paragraph especially supportive of its "future developments" section, which included the microwave straight-line glide path system being experimented on by the army and

MIT.[49] If we interpret Arnold's memo as referring to the microwave system, then there appears to be little ground for a charge of army resistance to the report. Bush's committee had, after all, validated the army's need for a straight glide path and had strongly preferred a microwave one.

The tone of Arnold's memo suggests that he had some unstated concerns, however, which may well have helped confuse General Watson. What those concerns may have been are suggested in the minutes of a conference held March 11 in Arnold's office. The administrator of Civil Aeronautics had requested the conference in order to find out why the Air Corps was resisting the report's recommendations. At that conference, an anonymous officer recorded, it was carefully explained to Hester that neither the Air Corps nor the Bureau of Aeronautics (Navy Department) had any objections to "ten purely experimental instrument landing systems, but felt that with the microwave system so far along, it would be undesirable to invest funds in radio equipment which would soon become obsolete."[50] The drafter reported that Hester had agreed that these were purely experimental and that the CAA would support the installation of a better system as soon as one was developed and satisfactory for general use.

Clearly, the Air Corps did not intend to adopt the CAA's system at all, as it preferred to wait the expected two years before a procurable microwave system was available. Arnold's concern was probably over the possibility that Hester was not negotiating in good faith. Arnold remembered clearly the Bureau of Air Commerce's adoption and near-simultaneous abandonment of the Hegenberger system. Many officers in the Air Corps had seen that as an act of betrayal. In this case, Arnold was likely concerned that the CAA might take the NAS's approval of the Indianapolis system as an excuse to deploy the system widely, making getting rid of it nearly impossible. Widespread commercial deployment of the system would almost inevitably force the Air Corps to adopt it, whether or not it satisfied their perceived needs.

The CAA did have larger plans for the Indianapolis system than the ten installations that NAS had approved, and it had budget authority for twenty-five in the 1940 and 1941 fiscal years. One CAA official had also suggested to the RTCA (with the army members present) that up to fifty might be deployed before a microwave system became available. That amounted to an investment of $1.25 million in 1940, if CAA's 1939 estimate of $25,000 per installation had been accurate, all of which would have to be abandoned when a microwave system was deployed in two or three years. Because Arnold complained bitterly in his autobiography about the shoestring budget of his Air Corps, it seems likely that he did not consider such a sum easily abandonable. Congress may well have balked at any CAA plan to replace the UHF system, too, as it was already unhappy at the cost of maintaining the Federal Airways System.[51] Arnold's concern with being stuck with an unsuitable system was probably well founded.

The March meeting adjourned with a promise by Arnold's office to draft a memo to General Watson to correct his misunderstanding of the Air Corps' position. Arnold submitted such a memo, which unfortunately no longer exists, but it was not enough to satisfy Watson. On April 3, the War Department chief of staff requested that Brigadier General Yount, Arnold's assistant chief of the Air Corps, go to the White House to explain the matter to General Watson in person. General Yount's conference with General Watson resulted in a request for yet another memo, this time signed by the secretary of war, which was duly generated and sent to Watson on April 13, 1940. Watson dispatched that memo to President Roosevelt, and Roosevelt attached his "OK. Execute" on May 2nd.[52] The RTCA's recommendation, changed only in emphasis by the NAS committee, had finally achieved the status of an official standard.

CONCLUSION

The process of constructing a standard landing system was made painful by several interrelated factors. Each organization involved in the negotiations had a particular set of technical demands, which mostly, but not completely, overlapped. Their technical demands, in turn, were based upon the types of aircraft that they intended to operate. For example, the navy's prewar focus on seaplanes made it in many ways the "odd man out" in these talks, playing a distinctly secondary role to the Air Corps and the CAA. The curved glide path was well suited to its seaplanes, and because no one else used flying boats, it was allowed to go its own way.[53] The navy's unusual technological suite imposed unique requirements on its choice of landing system. Since it did not need to share airfields with the army or airlines, it had no reason to conform with any standard, and no one else chose to take issue with its nonconformity.

The Air Corps and commercial airlines, on the other hand, recognized that their aircraft had to be able to use each other's airfields in case of emergency or war and agreed that a standard was necessary, but they disagreed on the timing of that standard. An "almost straight" glide path was temporarily acceptable to the airlines, until such time as the "ultimate" straight path was available. The airlines, and the government agency created to serve them, wanted an interim system sooner rather than a final system later. This was a matter of economics, as suggested earlier. A landing aid system, even one at only ten airports, was expected to pay for itself quickly though increasing the number of completed flights. Hence, the technical issue of glide path curvature was secondary to the airlines' economic interests. The CAA's goal was happy constituents, and the airlines had powerful friends in Congress to whom unhappiness could be expressed—and it was the following year. The CAA's support for a temporary solution, even one costing large sums, was therefore perfectly reasonable, especially since the CAA's investment in ground transmitters amounted to less than the airline investment in receivers.

For the Air Corps, the technical and economic issues suggested a different timetable. The Air Corps believed that its new big bombers needed a longer approach than that provided by the Indianapolis system, and it therefore wanted to proceed directly to the ultimate system, which it expected to be available at the same time that these aircraft reached full production. Without a war to provide the operational necessity to justify the expense of a temporary solution, the Air Corps' technological needs were reinforced by its own financial analysis. With ten times as many aircraft to equip as the scheduled air carriers possessed, the Air Corps faced an enormous investment in receivers with no visible return, as well as the additional burden of having to explain to a Congress still suspicious of military spending why such an investment would have to be scrapped in a mere three years. Focusing on the ultimate system was the most reasonable approach for the Air Corps, technically and economically. The differing needs of these organizations was merely exacerbated by the lack of a policy to guide approval of interdepartmental standards.

Finally, the agreement achieved closure within the community on the proper form that the solution to the blind landing problem should take. It had agreed on the tripartite NBS system, consisting of marker beacons, localizer, and a straight glide path. This represented less an exemplary artifact than an exemplary model upon which equipment and training could be based. That model, which places the information needed to land a plane directly in the cockpit, I call the pilot-control model. Only two years after Roosevelt approved the NAS report, that model was challenged by a new one based on a radically different principle of operation. The result was a political crisis.

CHAPTER FIVE

Instrument Landing Goes to War

President Roosevelt's May 1940 approval of the National Academy of Science's report on blind landing systems had established a "standard" interim system and a research program to develop a final system. That approval allowed the army and CAA to continue supporting Edward Bowles's microwave project while CAA began installation of the allowed ten copies of the Indianapolis system. But U.S. entry into World War II caused the aviation community to abandon its carefully negotiated agreement. The army, suddenly needing a system sooner than later, embarked on three different development programs to produce a better glide path, while also supporting an MIT Radiation Lab project to produce a radar-based system. That radar project, called ground-controlled approach, is the subject of Chapter 6.

The three glide path projects the Army Air Forces embarked on were a straight very high frequency (VHF) glide path, for which it contracted with International Telephone Development Corporation, CAA's contractor; a ten-centimeter continuous wave (CW) microwave system based on Bowles's work, carried out under a Signal Corps contract to Sperry Gyroscope Company; and a ten-centimeter pulsed glide path (PGP), which the MIT Radiation Lab began on its own initiative. The VHF system, called SCS-51 during the war, was used by the U.S. Army Air Forces in Europe, North America, and in the Pacific theater by transport aircraft and bombers. It was functionally identical to CAA's postwar instrument landing system (ILS), and differed from the prewar Indianapolis system only in being portable and in using a straight, equi-signal, VHF glide path.[1] The temporary system thus became a permanent fixture of the aviation infrastructure, while both microwave systems—the wave of the future in 1940—have vanished into the past.

There is no other explanation for the entrenchment of VHF at the expense of microwaves other than the war itself. Entry into the war caused the United States to provide the AAF far greater resources, both financial and scientific, with which to pursue technological development. At the same time, those projects were constrained by immediate needs. The AAF could no longer wait for the future. It needed a landing aid immediately to avoid the huge weather-related losses that its new ally, the Royal Air Force, was already experiencing. The VHF system was the obvious choice if its glide path could be straightened quickly. At the same time, Sperry Gyroscope's microwave project ran afoul of the MIT Radiation Lab's belief in the superiority of its own work. In an odd paradox, the lab's devotion to magnetrons prevented the production of a microwave glide path during the war, leaving the field to the "low-tech" VHF system.

ENTRENCHING THE LOW-TECH SOLUTION, PART I: SCS-51 AND WORLD WAR II

In the two years following Franklin D. Roosevelt's blessing of the National Academy of Sciences report, CAA's Indianapolis system reached its nadir and was then suddenly, and permanently, rescued. A mistaken cost estimate nearly undid CAA's program, while the pressing demands of a world war provided a golden opportunity for CAA to salvage its shipwrecked plan and, in a stunning reversal, forced the Army Air Forces to adopt it too. The war thus put the allegedly temporary Indianapolis system firmly on the road to permanency.

In 1940, CAA found that it had greatly underestimated the Indianapolis system's cost. CAA had based its budget request for funding of the ten systems based on estimates made by its own experiment station personnel. They, in turn, had assumed that production units could be had at the same cost as the prototype contract, $25,000. CAA had therefore made a $250,000 request in its 1940 budget estimate, submitted to the Bureau of the Budget in late 1938. It received the full amount and found it was not nearly enough when it opened the three bids submitted by Air-Track, Bendix, and the International Telephone Development Company (ITD). Instead of ten units, it had only sufficient funds for four. Although CAA did let the contract to ITD, which was the low bidder, realization that only four sets would be installed during 1940 threw its plans into disarray. The airlines decided that four installations were too few to justify the cost of installing receivers in their planes and suddenly balked at CAA's plans. At least six installations were necessary to make back the cost of receivers before the whole system would be discarded in favor of a microwave one in three years or so, the airlines informed one of their Army Air Forces contacts.[2]

CAA's contractor also ran into engineering difficulties on top of the financial problems. The extensive modifications CAA had made to the original prototype meant that it had to be reverse engineered in order to make accurate drawings that could be used as a basis for manufacturing. The glide path transmitter had undergone the most extensive changes and required the most engineering work. That, in turn, consumed time, further delaying production and increasing costs. CAA did not get any equipment before 1941 and still had not managed to convince the airlines to buy receivers.[3]

CAA's mistaken estimate threatened to kill the UHF system before it reached deployment. With the temporary Indianapolis system delayed for at least a year while the microwave system work forged ahead, the airlines believed it made little sense to invest in UHF receivers. The airlines therefore abandoned the system they had been pushing for years, causing no little feeling of betrayal at CAA. Nevertheless, CAA went ahead with its purchase of the four installations, while trying to convince the airlines to return to the fold.

While CAA was having problems with its system, events in the larger world convinced the Army Air Corps that it could not afford to wait for a microwave system to emerge from Sperry Gyroscope's production facilities in a couple of years. The contacts forged between the Royal Air Force and Army Air Forces during 1941 brought home just how severe a problem lack of a blind landing system could be. Although Bomber Command's statistics do not break out landing accidents, it lost hundreds of aircraft to noncombat causes during 1939–1941. Shortly before the war, the RAF had arranged a license to produce the Lorenz system in England, under the name standard beam approach (SBA) system. Like its predecessor, the SBA system was unstable, and in 1942 physicist David Langmuir reported to Lee DuBridge at the MIT Radiation Lab that RAF pilots had stopped using it. Lack of an effective landing aid coupled with the extreme density of aircraft involved in landing several hundred bombers in an area of about a hundred square miles within forty-five minutes, it seems reasonable to infer that many of Bomber Command's noncombat losses were approach and landing accidents. Hap Arnold, who traveled to Britain in April 1941 for conferences with his RAF counterparts and an audience with the king, could not have been unaware of RAF's problems.[4]

The army thus continued to fund Sperry's continuous wave glide path project, while dropping the microwave localizer and marker beacons. The army intended to graft a microwave glide path onto the UHF CAA system, just as Bowles had recommended to Bush as the fastest approach to a useable system. With this "mix and match" approach to a landing aid, the ten-unit restriction was no longer necessary. With the approval of the army, therefore, RTCA recommended that the ten-unit restriction in the agreed-upon plan be dropped. This was intended to allow civil and military aviation to install and use individual elements of the approved system (i.e., localizer and marker beacons) while work continued on a variety of approaches to solving the glide path problem.[5] One could make a low approach using only the localizer and marker beacons, after all, and although this was not a perfect solution to the blind landing problem, it was very much better than nothing. A later RTCA decision added the Army Air Corps' compass locator stations as optional equipment; after the war, commercial pilots insisted that they be required because they made interception of the very narrow localizer beam much easier. The army's A-1 system hardware and CAA's system thus became completely integrated, on paper at least, by the end of 1941.

The Japanese attack on Pearl Harbor precipitated a massive installation program. With active prosecution of a strategic air war targeted for late 1942, the newly renamed Army Air Forces had to construct a ferry route up the East Coast to Newfoundland and thence across Iceland and into Scotland to get its new bombers to England. With CAA's system the only one immediately available, the AAF officially adopted all but the glide path part in December 1941. Recognizing that CAA was the expert on its own system, the secretary of war transferred more than $1 million to CAA to procure and install the marker beacons and localizer portions of its system throughout the United States, based on a prioritized list provided by AAF.[6] Glide path transmitters were to be provided by the army once it had developed one to its liking.

With only the glide path standing in the way of a complete system, and flush with suddenly vast development resources, the army pursued three different glide path projects in the hope that at least one might work out, and do so quickly. It continued supporting Sperry Gyroscope's continuous wave glide path project, based on Bowles's work, it entered into a contract with ITD for a 330 MHz equal signal glide path, and it approved a Radiation Laboratory proposal to build a pulsed glide path (PGP) based on the cavity magnetron. It also established a contract with ITD to build a completely portable and militarized version of CAA's system (minus the glide path, of course), which the company had begun to do with its own resources earlier in the year.[7]

The Signal Corps had taken over CAA's Indianapolis Experiment Station, its personnel, and its equipment, in early 1942 in order to more rapidly bring about an army version of the CAA's system. It brought aboard ITD, which had changed its name to International Telephone and Radio Manufacturing Company (ITRM), to do the production engineering work on a portable system that was to be designed around standard tube sets. The army intended the use of standardized, easily available tubes to prevent having the system's production delayed by parts shortages, which were already becoming common. It also improved the system's maintainability by eliminating the need for specialty tubes, which would not normally be stocked in the supply system. These were important considerations in a wartime system.

The army still insisted on a straight glide path, however, and while ITRM's engineers reworked the localizer and marker beacons for portability and standard parts, they also had to deal with the army's demand for a long, straight glide path. They did this by raising the frequency, changing the antenna, and adapting the equal signal method used in CAA's localizer for use in the glide path.

CAA's glide path had operated on 93.7 MHz, while the localizer worked on 110 MHz. The two different frequencies required two transmitters, consuming more tubes and extra space. ITRM's engineers replaced the glide path transmitter with a frequency multiplier, which raised the localizer transmitter frequency to 330 MHz (a VHF frequency) for broadcast through the glide path antenna. This saved space and tubes, but more importantly, the higher frequency reduced the environmental sensitivity of the glide path while improving the predictability of its propagation. Predictability was important because the adoption of the equal signal method allowed the glide path to be controlled to some extent. (Figure 5.1 shows the propagation pattern of ITRM's glide path.) By broadcasting a 150-Hz-modulated signal from the upper antenna of the new glide path antenna system, and a 90-Hz-modulated signal from the lower antenna, the system created an overlapping series of lobes that could be adjusted relative to the ground by altering the relative intensity of the signals transmitted from the two antennas. A technician could thereby adjust the glide path to compensate for changing ground conditions.

The major drawback of this method is that it produced several possible glide paths, most of which were unflyable. Each of the 150-Hz lobes could appear as the correct glide path to a pilot. The possibility of intercepting the wrong glide path was reduced by procedure: the approach chart for each field directed pilots to be at a particular altitude when they reached the outer marker beacon, which corresponded to the altitude of the correct glide path at that point. (Figure 5.2 is the 1946 approach chart for Newark.) The chart's lower section shows an elevation view of the approach, specifying that an approaching aircraft be at 800 feet

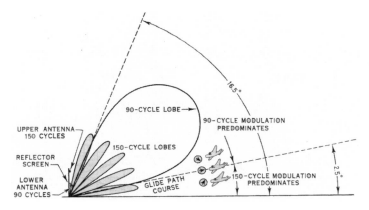

Figure 5.1. The SCS-51 glide path antenna radiation pattern. Only the lowest 150-cycle lobe (light gray) is a useable approach angle for most aircraft. Because the two antennae were independently excited, by adjusting the amount of energy supplied to them, an operator could adjust the angle that the 150-Hz lobes made with the ground to a limited extent. This helped compensate for changes in moisture. A monitoring device sounded an alarm if the lobe patterns deviated more than a preestablished amount from the desired pattern. M. E. Montgomery, "Latest Type AAF Blind Landing Equipment," *Electronic Industries* (January 1945): 101.

when passing through the outer marker beacon at Metuchen, thirteen miles from the field. An aircraft at that altitude, and on the proper course, would intercept the correct glide path.

The use of systematic procedures to reduce the probability of accidents was already common practice in aviation, although its application to airfield approaches had not been standardized and codified, even for individual airfields, before World War II. Chaotic conditions due to the vast increase in air traffic during 1942 forced the AAF and CAA to establish a joint board to standardize and publish the approach procedure for each major airfield in the United States so that all pilots near an airport would be following the same procedure, reducing the probability of collisions while also increasing airfield handling capacity.[8] That standardization was also necessary for successful use of the new glide path; however, the army's newly completed landing system did not drive the standardization process. Standardization began as a response to vastly increased traffic.

International Telephone and Radio's glide path was placed into a competitive flyoff in late 1942 against the two microwave systems, where it performed well enough to win a Signal Corps order for 350 units, which, combined with the localizer and marker beacons, were to bear the designation SCS-51 (see Figure 5.3).

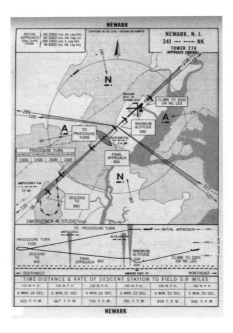

Figure 5.2. The standardized 1946 approach chart for Newark airport. The top section shows the approach as it appears from above, while the smaller middle section shows the approach's vertical aspect. The bottom section provides rate of descent for several landing speeds, and the length of time a descent will last at that speed and rate of descent from the inner marker (Elizabeth, 0.9 miles from the field). The minimum ceiling at this airport was 500 feet, as the chart shows. The approach manual required that pilots unable to see the ground from 500 feet immediately climb back to 2,000 feet and request to "go around" for another try. *Army Air Forces Instrument Letdown Procedures,* 1 September 1946, p. 91.

Early production models were subjected to extensive testing at various locations in the United States during 1943, including locations in Alaska. There, the joint air forces/navy command found that it worked well enough to recommend permanent adoption, and largely on the strength of the navy commander's recommendation, the navy agreed to accept SCS-51 as its own new standard later that year.[9] The Army Air Forces established SCS-51 as its standard instrument approach in 1943 as well, despite the system's unsuitability for small aircraft.

The portable SCS-51 was deployed domestically first, because a low priority rating hamstrung CAA's program to make permanent installations of similar equipment. Equipment for domestic use was automatically placed low on the priority list, while equipment intended for use in the combat theaters was given higher

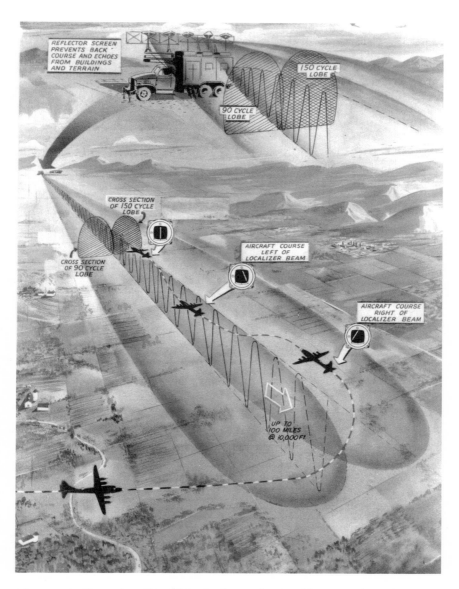

Figure 5.3. An illustration of the SCS-51 localizer path. Top of the image shows the localizer truck. The correct "path" is the surface of equivalent signal—when the signal strengths of the 90-cycle and 150-cycle lobes are equal. National Air and Space Museum (NASM A-4974-A), Smithsonian Institution.

priority. Because SCS-51 was designated vital tactical equipment, it received one of the highest priorities. By November 1943, less than a year after its selection, SCS-51 had been installed at fields along the northeast all-weather ferry and air cargo route to Europe, from New York (Mitchell Field and Newark), through Westover Field in Massachusetts; Presque Isle, Maine; and Harmon Field, Newfoundland. The vast number of receivers needed to equip all of the AAF aircraft to use the system, however, took many months despite the high priority. High volume production of receivers took until mid-1944 to achieve, with new production aircraft, particularly the B-29s, getting most of the first batches.[10] Hap Arnold intended to deploy SCS-51 to the Pacific theater with the early Twentieth Air Force B-29 units, although I have found no records attesting to specific locations. Administrative records dealing with the Pacific theater are in general far less available than those for the European theater, and to describe SCS-51's operational record, we will have to rely on Eighth Air Force experiences with it.

All U.S. air bases in Britain had been built with RAF equipment, and they were equipped with SBA systems. Accordingly, Eighth Bomber Command had adopted RAF's system as its landing aid when it first arrived in Britain. Bringing in a new system made little sense, and in any case, the AAF did not have one available. Eighth Bomber Command ordered several thousand sets of SBA equipment in 1942, which were to be shipped back to the United States and installed in the aircraft at the factories. These were never built. Although it did receive small numbers of SBA receivers, the Eighth never obtained enough to equip most of its aircraft, and therefore had no blind landing system at all until late 1944.

Therefore, like the RAF, which by 1943 had begun abandoning SBA due to lack of receivers, Eighth Bomber Command did not fly missions if the ceiling for the return flight was expected to be less than 500 feet, and if forecasts turned out to be wrong, bomb groups broke up into squadrons, which descended to one hundred feet or so above the English Channel and then flew by visual landmark back to their fields. Various bomb groups worked out their own approach procedures based on their home field and aircraft equipment. Aircraft equipped with Gee, a navigation system designed primarily for bombing, could make approaches using that system, with the plane's navigator feeding the pilot directions.[11] Aircraft without Gee sometimes could use compass locator stations called "slashers," and intended as raid-forming beacons to shoot radio compass approaches to their home fields, much like the procedure for the army's Hegenberger system. In truly blind conditions, airfields also stationed men with flare guns at either end of the landing field. The soldier at the approach end fired a green flare when he heard a plane passing overhead; if the man at the departure end heard a plane flying overhead, he released a red flare. The gun crews in the planes were expected to watch for the flares and tell the pilot whether to land or not. There were, therefore, several field solutions to the blind approach problem, although all were dangerous for novice pilots.

With the establishment of SCS-51s along the ferry route in late 1943, the equipment started to earn a reputation for itself, and Gen. Carl Spaatz, commander of U.S. air forces in Europe (USSTAF), requested that a unit be sent to the United Kingdom for trial there. The equipment arrived in late January 1944 and in early February a two-week long series of tests took place at Defford, attended by senior commanders from Eighth and Ninth Air Forces, RAF's various commands, and the joint chiefs. Based on these tests, the joint chiefs adopted SCS-51 for use by both RAF and USAAF in the European theater immediately. Initially, thirty sets of ground equipment were required, with enough receiver sets to entirely equip the Eighth and Ninth Bomber Commands' aircraft.[12]

The major selling point for European theater commanders was SCS-51's easy adaptability to automatic control. At the tests at Defford, Signal Corps Lt. Col. Francis L. Moseley, formerly an engineer for Sperry Gyroscope, had demonstrated a device that converted the system's localizer receiver output into a signal that the aircraft's autopilot could use. Although the device was not intended to land the plane, it significantly reduced the pilot's workload during the approach by doing most of the flying. Essentially, the pilot's job during an approach became monitoring the system's performance, dealing with the throttles, and taking over when the field was in sight. Spaatz liked the idea enough to demand that the experimental device be procured immediately as an integral part of the system. The demand for this "automatic coupler" was driven, in part, by the legacy of the failed SBA system, which had left pilots distrustful of the very similar SCS-51.[13] The automated approach removed much of the mental stress that blind landings entailed, while removing many of the skill requirements (and therefore training requirements) as well. In short, it fit the Army Air Force's wartime need for a system that its relatively inexperienced pilots could use.

Spaatz's demand for the automatic landing coupler caused the Signal Corps to embark upon a crash program to develop versions of it compatible with the variety of automatic pilots in use, including the Honeywell C-1 and Sperry A-5, which ran the duration of the war. The device never made it to combat theaters, however, because it could not be produced in time. Similarly, the Eighth Bomber Command did not receive enough glide path receivers for all of its aircraft until

January 1945, despite the receiver's design having been completed in late 1943.[14] Even high priority equipment could not be manufactured so quickly in such large volumes.

USSTAF headquarters kept up the pressure, however, complaining in September 1944 that lack of glide path receivers was preventing C-47s of Air Transport Command from keeping supplies moving to the front, and that the automatic coupler was vitally needed.[15] By early 1945, enough glide path receivers had reached Europe to allow equipping all bombardment and transport aircraft with them. This left the AAF with a training problem, since its pilots had not been trained to use the system.

The AAF had attempted to establish an instrument landing training program in 1942 using sets of YB equipment borrowed from the U.S. Navy.[16] Because the YB system was very similar in operation to the SBA system pilots deploying to Europe were expected to use, pilots would be able to gain familiarity with the SBA's landing procedures while still at the army's training schools. The idea failed in application, however, for the same reason the navy abandoned the system the following year: it was simply too unstable, especially for training use. The training schools quickly dropped the idea, and pilots deployed to Europe with no training in instrument landing.

Instrument landing training therefore had to be done in the combat theater. To accomplish that, USSTAF pressed into service the Link trainer, which could be equipped to simulate instrument landing the same way it already simulated instrument flying.[17] A cross-pointer instrument in the trainer's cockpit was manipulated remotely by the trainer operator. Pilots in the trainers then used the indications to fly the trainer with, as a pilot would if the trainer were a real aircraft. Some pilots reported that they received no training at all, while others report having been able to practice on the Link equipment between missions. Hence the opportunity to train on the Link equipment was clearly not available to all pilots.

To help ameliorate the training problem, the AAF's training manual suggested that pilots practice on the real system as much as possible, by making their approaches "under the hood" even in good weather, with the copilot completing the landing visually. With only thirty SCS-51s in the European theater, however, most fields did not have one to practice on.[18] Instead, the sites for SCS-51 installation were usually chosen so that the equipped field could serve as an emergency field for several nearby bases. That was necessary due to the dense spacing of fields in England, combined with the limited number of channels available to SCS-51. Although more equipment could have been set up, the mutual interference would have rendered the sites useless. Hence training for SCS-51 proved a severe problem for the AAF, which had thousands of crews to train in Britain.

Even without the automatic approach coupler, however, and with a limited training program, pilots who had access to SCS-51 seem to have appreciated it. One pilot remembered the glide path was the most satisfying part of the system because it eliminated the altitude uncertainty remaining in the barometric altimeter. It greatly reduced the number of "go arounds" because aircraft broke through the clouds in consistently better position for landing. The AAF had long ago given up on the idea of routine blind landings, and this pilot's recollection is exactly what the AAF's leaders had hoped to achieve.

World War II, therefore, rescued the Indianapolis system from an oblivion virtually guaranteed by CAA's mistaken cost estimate and the resulting defection of the airlines. The war also established the system's reputation by demonstrating that a different VHF glide path would work. Finally, by instigating widespread deployment of this hybrid system, the war had made it easy to justify adopting the system permanently. Vannevar Bush had warned against the dangers of deploying a system too early, thus making far more difficult its replacement by a superior system, and as we will see, his foresight had been correct.

A BRIEF INTERLUDE: DEMISE OF THE HIGH-TECH SOLUTION

The army did not stop supporting Sperry Gyroscope's work to construct a microwave glide path when it decided to purchase International Telephone's VHF one. Instead, it pursued both projects. Success of the VHF glide path was therefore not solely responsible for the failure of the microwave one. The MIT Radiation Lab undid Sperry's work by challenging Sperry's glide path with one of its own.

Sperry Gyroscope, which had long been a contractor for both the U.S. Army and U.S. Navy for navigation equipment, entered microwave work when it contracted with Stanford's physics department to fund the development of the klystron into a commercial product. Chapter 4 details Edward Bowles's work on a prototype forty-centimeter blind landing system at MIT, and the Signal Corps' award of a contract to Sperry to build a complete blind landing system based on Bowles's work. That system proved to have one major problem: size. The radiator horns needed to broadcast efficiently at forty centimeters were too big to be portable, and were also a significant collision hazard for aircraft. The obvious solution was to use a shorter wavelength, which would require smaller radiators.

Figure 5.4. The Sperry Gyroscope microwave glide path transmitter. Note that the original horn radiators have been replaced by what is essentially a vertical "slice" through them. With microwave transmission, the width of the beam is inversely related to the transmitting antenna's dimensions. In this case, the antenna produced a pattern that was wide horizontally but very narrow vertically. Courtesy of Hagley Museum and Library.

The Signal Corps, CAA, and Sperry chose to stop work on a forty-centimeter system, and put their development effort into a ten-centimeter system.[19] They began testing that system in 1942 (see Figure 5.4).

Stanford's researchers and Sperry's engineers had always intended to use klystrons for continuous wave transmission, and the Sperry Gyroscope ten-centimeter glide path was built in that tradition. In two works examining the development of the two early forms of radio transmission, radiotelegraphy (based on interrupted wave signaling) and radiotelephony (based on continuous wave transmission), Hugh Aitken has argued that the two different forms of transmission represented two different intellectual traditions.[20] Practitioners of one could not easily adapt to the other, because the mental tools needed to work in one tradition were inappropriate for the other. Although Aitken relied on radiotelephony narrowly defined—we would call it radio broadcasting—continuous wave transmission with, or without, voice modulation, utilizes the same techniques. Sperry's engineers thus worked within a tradition, or technological frame, of continuous wave transmission. When faced with pulsed transmissions, they adapted poorly.

The pulsed system, which was supposed to supersede Sperry's work, was a product of a new organization, the MIT Radiation Laboratory. It had been founded in 1940 by Vannevar Bush's National Defense Research Committee to pursue the development of microwave techniques for use in the war effort.[21] The lab's foundation had been provided by the arrival of British physicist Henry Tizard in the United States with a new device for generating microwaves: the cavity magnetron. The magnetron, like the klystron, relied upon electrically resonant cavities to produce microwaves. The internal structures of the two tubes, however, was completely different, with the result that the magnetron was better suited to providing high power output if operated intermittently. Because higher power translated directly into longer range, clearly a benefit for weapons systems, the Radiation Lab focused exclusively on designing equipment for pulsed use, and the lab's entrant into the glide path competition was no different. Instead of broadcasting continuously, the Pulsed Glide Path (PGP) transmitted a train of pulses that aircraft could receive.

The lab decided to produce a glide path after Vannevar Bush assembled a committee in November 1941 to revisit the blind landing system progress obtained since his investigation two years earlier. Alfred Loomis, a former industrialist turned amateur physicist, chaired the committee. The group reinvestigated the blind landing work being done by CAA, army, navy, and private companies around the country. It reviewed nine systems in all: Hegenberger, Air Track, the CAA's VHF system, the CAA-MIT microwave system, Sperry's microwave system, PGP, GCA, the new 330-MHz glide path being made by ITRM for the army, and the Lorenz system. The committee recommended that the lab develop the PGP for production due to the rapid accumulation of pulsed techniques and the increasing availability of equipment designed for pulsed operations.[22] Perhaps unsurprisingly, it thought little of the possibilities for any of the non–Radiation

Lab projects, however, and other documents suggest that it thought Sperry's system too complex to work. Hence, the lab decided to make a full-scale push toward getting the PGP adopted for production by the Signal Corps.

The PGP project had begun informally earlier in 1941, administered under the lab's Group 73, the "Landing Group." J. S. Buck was the project engineer. His goal was to produce a glide path operated on three centimeters to minimize the radiators' size. The PGP, which the lab developed in partnership with General Motors' Delco Electronics subsidiary, operated by sending a train of pulses through a horn radiator, producing a narrow beam. The receiver employed an averaging circuit so that it provided a continuous indication on a cross-pointer type instrument, making the pulses invisible to the pilot. Unfortunately, there is practically no other information available on the PGP project. I have been unable to locate either photographs or a detailed description of its operation. Fortunately, there are documents relating to flight testing and the acceptance of the two microwave systems still available, despite the lack of technical documentation.

A series of tests held in late 1942 between PGP, the Sperry continuous wave system, and the SCS-51 system from the ITRM held at Pittsburgh, Cincinnati, Indianapolis, and Wright Field, resulted in an army decision to buy both PGP and the ITRM system. The Sperry continuous wave system was to be abandoned. The reports do not make clear why PGP was chosen over Sperry's system, as each appeared to perform equally well (or poorly, as neither put in particularly encouraging performances.) The evidence suggests, however, that the AAF had already come to rely heavily on the advice of the Radiation Lab's physicists in its electronics procurement, and it is very clear that the Radiation Lab supported its own program over Sperry's.[23] The Signal Corps chose to assign production of PGP to Sperry Gyroscope, a decision apparently based on an analysis that suggested Delco Electronics did not have sufficient production resources to manufacture PGP in addition to its other obligations. Sperry, in turn, never put PGP into production.

PGP did not reach production for two reasons, related directly to the company's previous work with the klystron and Bowles's continuous wave glide path. Sperry's engineers were not experienced in "pulse techniques," as the Radiation Lab's project supervisor put it.[24] Sperry's engineers were unable, or at least very unwilling, to adopt pulse techniques over the continuous wave design that they had spent several years developing. The company's managers also preferred to leave their engineers assigned to Sperry's own continuous wave system. The Radiation Lab's official historian, Henry Guerlac, attributes the nonproduction of PGP to Sperry's having lost interest in the project.[25] It seems fair to say, however, that Sperry was never very interested in the first place. Sperry had spent a great deal of time, energy, and money developing the klystron, upon which the continuous wave system was based and to which it also owned the rights. It could not recoup that investment if the klystron were supplanted by the cavity magnetron. The PGP was not in Sperry's long-range financial interests, any more than it was within the realm of Sperry's existing technical experience. Hence, Sperry's management chose to keep its engineers at work on the klystron-based continuous wave glide path throughout the war. This combination of lack of expertise and lack of management interest spelled the end of PGP, and ultimately of microwave glide paths.

The Radiation Lab's intrusion into microwave glide paths thus prevented the production of a microwave glide path during the war, but Sperry continued to work its microwave system, hoping that it could convince the AAF and CAA to replace the temporary VHF system. In 1946, it launched a substantial marketing effort to get its microwave system adopted as the U.S. and international standard. Yet the substantial number of SCS-51 and fixed CAA ILS systems installed during the war, and the tens of thousands of receivers for it produced during the war, proved to be an insurmountable obstacle. To suggest how fully the war had biased the aviation community against a rapid replacement of the "temporary" system, we examine the decision of the infant Provisional International Aviation Organization (PICAO) to rely upon the SCS-51 as the basis for an international landing aids standard. That decision served as the next nail in the microwave glide path's coffin.

ENTRENCHING THE LOW-TECH SOLUTION, PART II: PICAO

The Provisional International Civil Aviation Organization had been formed as the result of a series of conferences between Britain, the United States, Canada, and a host of newly restored governments, governments in exile, and governments of the few noncombatant states in late 1944 and early 1945.[26] PICAO's function was to establish the framework for a permanent organization to regulate international civil aviation. One of the powers granted to the organization by the member states was the ability to set technical standards for navigation and communication equipment, so that aircraft flying between nations did not have to carry different equipment for each of its destinations. The delegations all recognized the financial absurdity of that situation.

The problem of landing aids was only one of a number of technical issues that PICAO needed to resolve and probably the least controversial. The real "battle between systems" at the international level took place over short- and medium-range air navigation systems and was between the British delegation, led by the famous radar physicist Robert Watson-Watt, and the American group, led by career bureaucrat Charles Stanton, who had been administrator of civil aeronautics until fired by President Roosevelt in 1944.[27] This fight suggests some of the reasoning that likely underlay the less controversial, and therefore less well-documented, landing aids decision.

Watson-Watt demanded the adoption of the wartime Gee system as the future international short- and medium-range air navigation system. Gee operated via a series of ground stations laid out in a network. These stations generated time-based signals that an airborne receiver detected and decoded as a navigational grid, which could then be used by a navigator to determine the receiving aircraft's position. It was operationally similar to the long-range Loran system but more precise. Its major advantage, according to Watson-Watt, was that it allowed aircraft to operate anywhere within the area of broadcast coverage, meaning that aircraft were not confined to specific, narrow "airways" like those provided by the four-course ranges in the United States. This, he contended, would allow higher traffic densities than an airway-based system, which he believed would be necessary to serve Europe's dense population.[28]

Stanton, in contrast, promoted the U.S. airways model. Instead of producing a network of lines from which to derive a position, the U.S. system of radio ranges simply produced a line of bearing. Pilots then simply flew from range to range along well-defined airways. These airways were essentially highways in the sky. Fliers could leave the airways, but outside them pilots had no effective navigational references other than landmarks on the ground. Initially, the ranges were the four-course type discussed in Chapter 3, but during World War II a new sort of omnidirectional range had been devised that provided a reasonably accurate course in all directions. This was called a visual omni-range (VOR), the *visual* meaning that the information was displayed to the pilot on an instrument instead of aurally. Distance to the range was to be provided by another wartime innovation, called Distance Measuring Equipment (DME). The great advantage of this VOR-based system was that it was relatively simple and inexpensive, and Stanton openly ridiculed Watson-Watt over the cost of installing and operating the hundreds of short-range Gee transmitters necessary to cover North America.[29] VOR's biggest disadvantage, Watson-Watt correctly noted, was that it confined fliers to specific routes, greatly reducing the volume of airspace useable by instrument fliers.

Neither man presented a rigorous argument or substantial data to back up his claims, however, and the delegates chose the U.S. system. Most nations simply took over the airways stations that the U.S. Military Air Transport Service had installed to facilitate its worldwide wartime operations, and later expanded upon them.[30] It made little sense immediately after the war for nations to spend a lot of money on navigational aids when a system sufficient for their immediate, basic needs had already been installed.

A similar sort of reasoning made the adoption of the USAAF's SCS-51 as the international standard completely uncontroversial. SCS-51 had also already been installed, again by the U.S. Military Air Transport Service, at a number of major European airfields. It made sense to adopt a system that was available immediately and relatively cheap. New installations, at $70,000 apiece, also seemed a less budget-busting solution than the other technologies presented during demonstrations held in late 1946 for PICAO's benefit, which included the Sperry microwave system and the MIT Radiation Laboratory's Ground Controlled Approach system, expected to cost $200,000.

Further, as the Belgian delegate pointed out, other nations did not want to be forced to buy all of their equipment from a single source, and many countries wanted to be able to manufacture it for themselves.[31] He raised this as a condition for the acceptance of all PICAO standard equipments, and both Watson-Watt and Stanton quickly agreed. The chosen standard thus could not be proprietary or contain military secrets. It also had to be fairly easy to manufacture, which was manifestly not true for microwave equipment.

The most widely reported reason for PICAO's selection of ILS as the international standard, however, was its ease of adaptability to automatic landing. The automatic approach coupler that General Spaatz had demanded for Eighth Air Force in 1944 was very important to the plans of PICAO's technical committees, which sought greater automation of flying overall. Why they sought to automate landings was not made explicit, but it seems reasonable to assume that their goal was the reduction of missed approaches and landing accidents by elimination of "pilot error." No one seems to have expected the elimination of pilots themselves, since the control of a plane's attitude during landing, which was all the approach coupler did, was only one part of pilots' jobs. Landing was the highest workload phase of a flight, and the delegates no doubt believed that automating part of that workload would result in safer flights and improve the all-important regularity of service. When the first president of ICAO, American aeronautical engineer Edward P. Warner, discussed the technical work being done by the organization, he focused on the need for automation to reduce workload and errors.[32]

PICAO's delegates thus had several good reasons to adopt the low-tech SCS-

51 over its competition. They had the evidence of its wartime performance, they were able to fly the systems during demonstrations held in England and at Indianapolis, and they had specific goals of economy, national self-interest, and automaticity to help sway their decisions. The war, however, had provided the conditions that informed the choices of PICAO's national delegations.

CONCLUSION

From a scenic overview of the technoscape of aviation dated 1940, the future of Bowles's microwave work seemed assured. Microwaves were clearly the future. World War II altered that technoscape of aviation dramatically, however, by driving the Army Air Forces to adopt the CAA's previously "inadequate" system as a "good enough" expedient to support its wartime operations. At the same time, by causing Winston Churchill to dispatch Henry Tizard to the United States with one of the United Kingdom's most valuable possessions, the cavity magnetron, the war established the basis of the MIT Radiation Lab, and thus of the Pulsed Glide Path system. Without the pressing needs generated by the war, the AAF would not have contracted with International Telephone and Radio for the VHF glide path, which the Civil Aeronautics Administration incorporated into its postwar ILS, and the Radiation Lab would never have been founded to produce its challenger to Bowles's work. In sum, neither challenger to the Bowles-MIT-Sperry Gyroscope continuous wave system would have existed. Hap Arnold would have adopted Sperry's glide path when it was ready, and the airlines, as their sudden reluctance to adopt the CAA's VHF system in 1940 suggests, would happily have followed the AAF's lead. Without the war, finally, the conditions under which PICAO was established and made its technological choices would also not have existed. The pressing demands of World War II, then, radically altered the outcome of the landing aids development process.

The exigencies of war did not merely reverse the fortunes of the two 1940 glide path projects, however. The MIT Radiation Lab produced yet another challenger to the CAA's system, and this one was not merely a different way to produce a glide path. Instead, this radar-based system replaced the entire model pioneered by the National Bureau of Standards system. Instead of using radio beams to activate a cockpit instrument, the ground-controlled approach system used radar to inform operators on the ground of a plane's position. This system thus challenged the community's very conception of how a landing system should work, shattering the consensus that the National Bureau of Standard's pilot-centered model was the one best way for a landing aid to function.

CHAPTER SIX

The Intrusion of Newcomers

The blind landing systems detailed thus far worked in accordance with a single model, for lack of a better word. All of them, whether based upon cables or radio transmissions, provided a signal or set of signals that directly operated one or more instruments in a plane's cockpit. They were a primitive automated remote control. Pilots translated the instruments' readings into information meaningful to them in accordance with their training and experience and then maneuvered their aircraft. Because pilots retained control of the information translation process, they also retained their own autonomy. They needed no help from people on the ground, although, of course, they required a great deal of assistance from devices on the ground.

The preceding several chapters have shown a process of building a consensus within the aviation community toward what was less an exemplary artifact than a preferred model of a solution. By 1940, every group involved in the process of building a blind landing system had accepted that a complete system had to include marker beacons, a localizer, a straight glide path, and a set of receivers in the plane. No other possible solution was under serious consideration. This was the National Bureau of Standards' model, and it belonged to a category of systems that functioned under what I called in Chapter 4 the *pilot control* model.

It was not the only way to build a blind landing system, however. Three years after the National Academy of Science's (NAS) report was so painfully hammered out, a radically different way of directing aircraft to a safe landing came into existence. The MIT Radiation Laboratory, in particular, future Nobel laureate Luis Alvarez, was responsible for challenging the pilot-control model. As a new orga-

nization, the Radiation Lab certainly was not a member of the dominant technological frame, and its leadership reinforced its outsider status by choosing to recruit from the ranks of nuclear physicists, as opposed to radio physicists and engineers. The Radiation Lab was thus able to create and introduce a radically new
model solution to the blind landing problem. This had far-reaching effects. The
old model left control of the aircraft in pilots' hands, where it had always been;
the new system placed control on the ground. Not everyone proved happy with
that sort of progress.

INVENTING A NEW MODEL:
THE CYCLOTRONEERS INVADE AVIATION

A month after President Franklin Roosevelt had signed the NAS report on the
status of blind landing systems, Vannevar Bush was back in his office, proposing
a scientific research organization devoted to developing new weapons, devices,
procedures, and materials for the war that both men believed the United States
could not avoid much longer. The result of that June 12, 1940, meeting with Roosevelt was the National Defense Research Committee, chaired by Bush, which
was to oversee research and development activities directed primarily at military
needs. The committee's original structure was poorly designed for the operation
of a large bureaucracy in official Washington, however, and in June 1941, Bush
obtained another executive order from Roosevelt submerging the committee under the Office of Scientific Research and Development (OSRD), once again, under Bush. The committee continued to exist but was restructured into numbered
divisions. Bush initially assigned all things microwave to Karl Compton's Division D, but after the reorganization, it fell under Division 14, and more particularly, the microwave committee chaired by Alfred Loomis.[1]

The spark that precipitated the MIT Radiation Laboratory was the arrival on
U.S. shores of Henry Tizard, a British physicist sent to secure American assistance in making use of several key British inventions. Tizard brought with him
the resonant cavity magnetron, which was capable of producing microwaves at
several orders of magnitude higher output than was the klystron that American
radars, and Bowles's microwave landing system, were based on. Demonstrated
to Alfred Loomis and Karl Compton on September 19, the magnetron's potential
convinced the two men that a central laboratory to develop applications for it was
necessary. Another meeting and demonstration held at Loomis's Tuxedo Park re-
treat in New York on October 12 and 13 cemented in his mind the need for an independent lab, and he convinced Bush. He and Bush chose MIT as the best site

for the lab, largely due to its existing microwave experience. Loomis and Bush badgered Frank Jewett into accepting their decision, and the three then lobbied Karl Compton, president of MIT, to accept it. He did so on October 17.[2]

With a site for the lab approved, the three men needed a staff. They decided to recruit from the ranks of nuclear physicists, since members of that discipline had developed extensive experience in dealing with high-frequency energy in the course of their cyclotron experiments. As historians John Heilbron and Robert Seidel put it, "cyclotroneers and their fellow travelers did not fear big projects, did not disdain to scrounge when necessary, did not insist on perfection or protocol. They were ideal people for crash programs."[3] With that understanding, Bush, Loomis, and Jewett turned to Ernest Lawrence to head the lab, but he chose to remain with his own work in Berkeley. Lawrence instead recruited Lee DuBridge, a former student of his who was then chair of the physics department at the University of Rochester, to fill the position. I. I. Rabi, of Columbia University, and F. Wheeler Loomis from the University of Illinois also signed on. For this story, however, the most important recruit Lawrence sent was another of his protégées, Luis Alvarez, who arrived in mid-November. Alvarez recruited a friend from his graduate student days at Chicago, George Comstock, who had taken up a teaching position, and slightly later, Lawrence Johnston, whom Alvarez had left in charge of his lab at Berkeley when he moved to MIT.[4]

Alvarez had joined Lawrence's staff in 1936 after completing his thesis under Arthur Compton at the University of Chicago, where he had learned to build things and to navigate a library, although he considered his education in physics there to have been poor. At Lawrence's lab, though, he learned the ropes of his chosen field. He finally read the physics he was supposed to have learned, and, more importantly, he learned to work in a team, which Chicago had frowned upon. Because Lawrence's cyclotron required a great deal of labor to maintain and operate, teamwork was necessary to keep the machine running. The machine required a six-man crew to run it each day, in shifts of two.[5] Alvarez blamed the regimen imposed by the machine for the lab's failure to make some of the fundamental discoveries in physics made elsewhere, but it served him in good stead at MIT, where his education in teamwork and hands-on operation of complicated equipment proved essential.

Physics was only one of his two careers, however. The other was aviation.[6] Both of his careers are important to this story. Alvarez had earned a private pilot's license in 1933, after three hours of instruction. He did not learn instrument flying until early 1942, when Aviation Chief Machinist's Mate Bruce Griffin taught him at the Squantum Naval Air Station in Massachusetts. He was therefore only a "contact rules" pilot when he conceived of his famous ground-controlled approach system, or GCA, in 1941. He was qualified to fly only within sight of the ground. His lack of instrument training meant that he had no mental commitments to the model upon which the other blind landing systems we have seen operated, although he knew their general principles. He was not set in the way it had always been done, nor was he a member of the dominant technological frame based upon the ideal of pilot control. He was free to imagine something completely different.

According to Johnston, Alvarez became interested in the blind landing problem while recuperating from gallbladder surgery in Rochester, Minnesota, in the summer of 1941. When he returned to MIT in August, he saw a demonstration of a gun-laying radar that gave him the basic idea for the GCA. The gun-laying radar, called XT-1 by the lab, was designed to fulfill one of the three primary tasks the microwave committee had assigned to the lab: an airborne intercept radar, the gun-laying radar, and a long-range air navigation system.[7] The XT-1, later given the official name SCR-584 (the SCR designates the equipment as Signal Corps Radio gear), was a truck-mounted, conical scan radar that automatically tracked an aircraft, and fed range, azimuth and elevation to an analog fire control computer that predicted the aircraft's future position, which could then be used to aim gunfire accurately.

While watching a test of the rooftop prototype XT-1, Alvarez realized that if radar could track an aircraft accurately enough to hit it with an artillery shell, it certainly could track one accurately enough to guide it to a runway. The basic idea, then, was to employ radar information to facilitate blind landings, instead of using a set of radio guide beams, as the various incarnations of the NBS system had. As it turned out, given the existing technologies, radar could be used in at least two ways to land aircraft, and both were used during World War II. One was Alvarez's ground-controlled approach; the other, a field innovation, was the blind approach beacon system (BABS).[8]

Alvarez's conception was to use the XT-1 to track an incoming plane and to use some modification of the analog gun director to provide an operator on the ground with range, bearing, and altitude information that he could then pass on to the plane's pilot via voice radio, which every military aircraft carried. The radar operator on the ground could therefore talk planes down. This was not a new idea. The navy already used a talk-down system aboard its aircraft carriers, with the "talking" done not by voice but by light wands in the hands of a crewman stationed on the ship's deck (the Landing Safety Enlisted, LSE). There was, of course, no radar involved in that process, but because the navy was already familiar with the basic principle of talking down, and in fact never did find a better way to facilitate carrier landings, it became interested in Alvarez's modification of the basic idea early on and after the war was its biggest proponent.[9]

Alvarez's idea got its first important boost from the battle against fog that the Royal Air Force (RAF) was fighting in Britain that fall. In some raids, Bomber Command was losing as many aircraft in landing accidents as it was over enemy territory due to the poor performance of its standard beam approach (SBA) system. Vannevar Bush therefore assembled another committee in November 1941 to revisit the blind landing system progress obtained since his investigation of two years before. Alfred Loomis chaired this committee, and Alvarez, who by then had been put in charge of Lab Division 7, the beacons group, was invited to be a member along with several representatives of the armed services. One outcome of this committee was pulsed glide path (PGP), as we saw in the previous chapter.[10]

The committee recommended that Alvarez pursue his ground-controlled approach idea quickly. Overlooked in all of the previous work in blind landing of aircraft had been the problem of small aircraft, which could not carry the receivers and instruments that the various versions of the NBS system required. This was particularly important for fighter aircraft (or pursuits, as they were called), which could afford neither the weight of the receivers nor the drag imposed by the two antennae without sacrificing performance. This had simply not been an issue before 1941, since without radar, fighters could not find targets in bad weather and therefore had no reason to be flying. With the development of radar, however, fighters could function in poor weather but only as long as visibility was still good enough to land. Because all of the equipment for GCA was on the ground, it imposed no weight or drag penalties, making it ideal for fighters and other small aircraft. GCA promised to complete a technological suite capable of allowing fighters all-weather operations.

The blessing of Loomis's committee permitted Alvarez to pursue GCA full time. He appointed Lawrence Johnston project engineer, and with George Comstock and David Griggs, a Harvard geophysicist and amateur pilot, set to work building an optical version of his GCA idea as a proof of method experiment. Two modified theodolites and a range-only radar were combined through a director device to produce a distance-to-landing indication, and two readouts of departure from an ideal glide path were established for the aircraft. During March 1942, the team put the optical equipment to work, directing Chief Machinists' Mate Griffin, flying a Grumman J2F "Duck," to safe landings via voice radio.[11]

The prototype XT-1 finally became available to the group during May 1942, and the navy invited Alvarez's group to Oceana Naval Air Station to try their landing idea out with the antiaircraft radar that had inspired Alvarez's idea in the first place. There, however, the project also nearly ended. The same reflection phenomenon that had caused so much grief for the NBS system nearly undid GCA. The system tried to track the aircraft's underground mirror image as well as the plane itself. Figure 6.1 shows the two types of mirror-image phenomena that XT-1 was susceptible to at low angles. As the figure indicates, the XT-1's antenna did not have a narrow enough reception pattern to keep the energy reflected from the ground from returning to the receiver at the low angle (3 degrees) above the ground that landing the aircraft required. The system had no way to distinguish the real airplane from its image, and its automatic tracking system hunted between the real and virtual planes, making it useless.[12] No simple modification to the XT-1 would turn it into a blind landing system. Alvarez had to completely redesign it.

Depressed over his misfortune, Alvarez met with Loomis in Boston. Alvarez later reminisced that during dinner at the Ritz Carlton, Loomis told him that "we both know that GCA is the only way planes will be blind-landed in this war, so we have to find some way to make it work."[13] Loomis forbade him from leaving the table until he had figured out a solution. The two agreed that the best approach was to replace the single XT antenna with two antennae, one vertical and one horizontal, which were to be mechanically scanned through narrow arcs. By making the antennae beaver-tail shaped, they narrowed the beam patterns, reducing the probability of reflections from anything but the target. These two antennae provided the short-range, highly precise indications needed to effect a landing, while a third search radar enabled the GCA's crew to direct aircraft into the narrow precision beams.

With Loomis's pledge of support to buy ten of the resulting systems with OSRD funds, Alvarez's group got to work assembling the Mark I GCA. Alvarez thought it would help speed production if the contractor sent engineers to the lab to learn as it was assembled, so OSRD asked the Signal Corps to suggest possible contractors. The Signal Corps originally selected Paramount Pictures to manufacture it, based apparently on the presence of a radio engineer named Homer Tasker on its staff. The movie company decided that it did not wish to enter the electronics business and agreed to loan Tasker to a Los Angeles company that made household radio receivers, Gilfillan Brothers.[14] Gilfillan thereby won the contract for the first ten units. Gilfillan dispatched Tasker and three others to the Radiation Lab in late 1942 to begin learning the system from its experts.

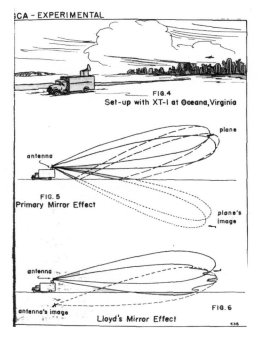

Figure 6.1. Two mirror effects prevented Alvarez's original idea for GCA from working effectively. In the "primary mirror effect," the aircraft reflected both radio energy directly from the antenna and energy which first bounced off the ground. The receiving antenna thus "saw" two aircraft, the actual plane, and its virtual reflection underground. In Lloyd's mirror effect, radio energy reflected off the ground interfered with radio energy following a direct path to the aircraft. The resultant of that interference is the dotted radar beam emanating from the "antenna's image." To the operator in the truck, the aircraft appeared displaced from its actual position. Lawrence Johnston, "Radiation Laboratory Report 438," file RB 334.8, box 1428, Chief Signal Officer Central Files, RG 111, National Archives.

The Mark I consisted of a gasoline-driven generator powering the radars mounted in one truck, with the antennae on its roof. A second truck contained the radar screens and the voice radio sets that controllers used to communicate directions to pilots. Both trucks were parked fifty feet to the left side of the runway in use, from the pilots' point of view. For production, Gilfillan's engineers replaced the second truck with a large panel trailer. The first production unit, named Mark II, is shown in Figure 6.2.[15] The engineers moved the antennae to the trailer, with the search antenna on upper right, elevation on the left corner,

Figure 6.2. AN/MPN-1 #1, at the Gilfillan plant in Los Angeles. The trailer contained the search radar (upper left), the precision azimuth antenna (lower center), and the precision elevation antenna (upper right), in addition to the operators' position (internal). The truck contained the power generators and air conditioning units. The individuals are not identified. Author's collection.

and azimuth in the front center of the trailer. The Mark II prime mover, the covered truck, contained a generator and air conditioning systems to keep the hundreds of vacuum tubes in the trailer from overheating.

Alvarez's group completed the Mark I in November 1942, and it was demonstrated to observers from the U.S. Army and the British Royal Air Force from mid-January to early February at East Boston airfield. A few days later, Gen. Harold McClelland, the director of Army Air Forces' (AAF) technical services, invited the group to demonstrate the system to more senior officers on Valentine's Day at National Airport. After several days of fixing wires and replacing tubes damaged by the long drive, Alvarez and his group successfully demonstrated the system to the AAF's satisfaction on the seventeenth.[16] Hap Arnold had anticipated success and had already requested that McClelland buy ten for further tests; by the end of March, the AAF had fifty-seven on order.

Much of the GCA testing before this official demonstration had been done at

Quonset Point Naval Air Station in Rhode Island, using navy planes and pilots, and was therefore already well known to the navy. The base commander had unstintingly praised the system, particularly after a January 1, 1943, incident in which a group of 3 PBYs was caught in a snowstorm and became lost. He had called Alvarez's group from the control tower and asked if they could bring the planes in. Alvarez talked them in, after gaining the trust of the pilots, who did not, according to Larry Johnston, even know radar existed. Alvarez was able to convince the pilots he could see them by talking them through various maneuvers before trying to land them. Once he had their trust, he was able to bring them in. The base commander reported the feat to the commander of Fleet Air at Quonset Point, and he soon requested that copies be procured by the navy.[17]

The initial orders from the AAF and the navy amounted to fifty-seven and twenty, respectively, in addition to the ten that OSRD purchased. The Army Signal Corps chose to let the AAF contracts to Gilfillan Brothers, while the navy chose to rely upon one of its traditional contractors, Bendix Radio. Both army and navy expected to receive their orders by the end of 1943; in fact, neither received any equipment before mid-1944, and neither service was able to provide them to combat theaters until 1945. GCA proved extremely difficult to get built because Alvarez was not as satisfied with it as the armed forces had been. Alvarez's group therefore spent the spring of 1943 converting the Mark I to the Mark II, which relied upon linear array antennae like those employed on one of Alvarez's other creations, the Eagle blind bombing radar. The advantage gained with this change was elimination of the mechanical scanning of the two precision antennae, which was a severe maintenance problem. The inertia of the large antennae destroyed the drive gearing quickly. The Mark II design also incorporated several changes to the radar displays, the most important of which was replacement of the B scan scopes with plan position indicator (PPI) displays, which showed the straight approach as a straight line, instead of a curve. With the improved presentation, the complex mechanical error readers that the Mark I had employed to mathematically transform the curved display into useful information about the actual straight approach path were no longer needed, reducing the number of operators and the complexity of the equipment.

By making these changes, Alvarez's group delayed achievement of a key milestone in production engineering, design lock-in. Without a fixed design to work with, the contractors could not make any headway toward establishing a manufacturing process for the system or obtaining parts. Although Tasker moved to the MIT Radiation Lab in late 1942 and became essentially part of Alvarez's team, he was not able to provide final specifics and had to frequently send changes back

to Gilfillan. The company then had to alter the orders it had placed with its sub-
contractors, delaying the process further. This was one consequence of Alvarez's
decision to change the design.

Another was that the contractors did not have the extensive diagrams they
needed to engineer system components because the Radiation Lab group had not
yet made them. Throughout 1943, with the design in flux, they did not finalize
the diagrams. The Signal Corps' Monmouth procurement district office, which
was responsible for overseeing the purchasing and production of GCA, criticized
the lab for refusing to turn over the prototype so that it could be reverse engineered to make the necessary drawings. Without drawings or the Mark I, the contractors had nothing to work with. Complete diagrams did not reach the contractors until October 1943, and Tasker had to have many of them redone. He
borrowed underemployed electrical engineers from the aircraft manufacturing
plants in the Los Angeles area to do that, but this delayed completion of the first
unit into February 1944.[18] Gilfillan did not complete the first ten OSRD units until May, after which full-scale production began.

The Mark I was also unavailable to the contractors because Alvarez still seems
to have thought that he needed to sell the system to the military brass, despite the
rapid placement of orders for GCA in March. He arranged to pack up the Mark I
in June and ship it to Britain for demonstration there, while Larry Johnston joined
Tasker at Gilfillan to engineer the Mark II. In Britain, Alvarez's team set up the
Mark I at Elsham Wolds, a bomber command base, for more demonstrations. The
British appointed an officer named Arthur C. Clarke to learn the equipment. Alvarez's group trained Clarke and his women (the RAF chose to use members of
its Women's Auxiliary Air Force to serve as crew under the supervision of male
regular officers) in maintaining and operating Mark I, and they in turn trained
other crews once Alvarez and his group left in August. The Mark I moved to
Davidstowe Moor in Cornwall, a coastal command base, after the Elsham Wolds
tests. It remained there as the centerpiece of the RAF's GCA school until mid-
1944, when its antenna drive gearing failed. The Mark I was cannibalized for
spares to use in the first three preproduction GCAs, which were given by the Radiation Lab to the British, as they had operators trained to use them.[19] The RAF
thus adopted GCA as quickly as had the AAF and the U.S. Navy.

In December 1944, George Comstock, who replaced Johnston as project engineer when Johnston left for Los Alamos in late 1943, reported that Gillfillan had
manufactured forty-seven GCA sets for the army, with fifty-five more on order.
Due to the extensive delays in production, the army had also placed an order for
one hundred more sets from Federal Telephone and Radio, only a few of which

had been produced by that date. Bendix had also produced only a few of the eighty sets for which the navy had contracted (sixty of which were intended for the British). Therefore, a total of something like sixty Mark II sets existed at the end of 1944, but none reached combat theaters until February 1945.

The bottleneck after 1944 was not equipment availability, then, but operator training. The Mark II GCA required a much larger crew than the original XT-1 inspired idea. Figure 6.3 shows the internal seating of the trailer. A traffic director operated the search set, while a plane selector designated a specific aircraft to land. The azimuth and elevation trackers operated the azimuth and elevation precision radars, respectively, and passed corrections to the controller, who directed the incoming pilot. Stuffed into a trailer, the five-person GCA crew required a good deal of teamwork among themselves to properly operate the system, in addition to the teamwork necessary between the controllers and pilots whom they were directing. The radar operators had to feed the controller the proper information quickly and accurately; the controllers had to reissue the information in the form of steering commands ("right 3 degrees," etc.) to the pilots; the pilots had to respond immediately to the controllers' commands, especially during the last few moments of a "blind" landing. Delay there could be fatal. GCA required a great deal of human coordination to work, and training with the actual equipment was the only way to forge such a team.

Because the Mark I had stayed in England, there was no equipment on which to train operators in the United States until mid-1944, when Gilfillan finally delivered the ten preproduction units. Nine of those went to serve as the basis of training schools, three each to the AAF, the RAF, and the U.S. Navy, while the tenth went to the Radiation Lab for further research. Training programs took time to establish, and more time to produce their first class. The AAF's training program, for example, was a twelve-week course and initially graduated thirty-three teams in early 1945.[20] By the German surrender in May, only eight GCAs with crews had reached AAF units in the European theater (all at fighter/bomber fields), and I can document only two in the Pacific theater by the Japanese surrender in August, although Radiation Lab documents indicate that ten crews had been allocated there.[21] A total of between ten and twenty GCA units, therefore, were in operation by the end of the war. This was a tiny fraction of all the airfields in use during the war, and GCA's contribution to the war effort was small. It was, however, an instant sensation among pilots.

The two great selling points for GCA's talk-down method were that it did not need special equipment in the aircraft and, more importantly, that it required no pilot training. Whereas all variations of the NBS system required extensive pilot

Figure 6.3. AN/MPN-1 GCA internal trailer seating. The traffic director on the right operated the search radar, while the plane selector assigned an aircraft to the controller for final approach. The traffic director maneuvered the selected plane (via voice radio to the pilot) into the precision radars' beams, whereupon the controller took over the approach. The azimuth and elevation trackers fed the controller information from the precision scopes to allow completion of the "blind" landing. Author's collection.

training to read and properly interpret its instruments and to gain the pilots' faith in the system itself and their ability to use it, the GCA only required the ability to respond to voice commands. Military pilots, at least, got that training beginning with their first day in flight school. Particularly important during the war was the human contact that GCA provided. Since crews often returned from missions with damaged aircraft, exhausted and sometimes with injuries to make their landings even more hazardous, being talked in by fellow soldiers was reassuring in a way that the NBS-type systems could never be. From the pilots' point of view, GCA was nearly miraculous. It could see them when they could see nothing and tell them where they were when they had no idea. The contrast with the NBS system, with its training requirements, its equipment needs, and its environmental vagaries, could not have been more striking.

Military pilots, and especially navy pilots, took to GCA immediately. In part, the system's basic requirements placed no demands on them other than following orders. Thinking was done by someone else; as RAF veteran Frank Griffiths put it, "any fool could use it."[22] Perhaps more importantly, however, GCA fit the culture of wartime military flying extremely well. Teamwork was required among bomber crews because no one could manage all of the plane's functions alone, any more than Lawrence's cyclotroneers had been able to individually manage their massive research instrument. Fighter and fighter-bomber pilots were also trained to work as teams with their squadron mates and wingmen, both as a matter of collective defense and to increase combat effectiveness. Hence, GCA's demand for teamwork between air and ground fit a model of behavior for which military pilots were already prepared.

This was the same model of behavior that Alvarez claims for Lawrence's laboratory, and to that esteemed place we should return to find GCA's genesis. While all of the inventors and innovators working under the NBS model attempted to reduce a blind landing system to one-man operation by a pilot from a lofty, bumpy, and moving perch—thus placing yet another burden on the pilot—Alvarez spread the work of a blind landing out among a group specially trained for it. Just as teamwork allowed Lawrence's cyclotron to keep going, teamwork powered GCA. Alvarez certainly did not learn the benefits of teamwork at Chicago, which he described as a solitary experience. He learned it from his experiences at Berkeley and built it into his invention at MIT. We must therefore place some of the praise for GCA—and some blame, for not everyone loved it—squarely at the feet of the cycolotroneers.

CONCLUSION

The MIT Radiation Lab closed down shortly after the war, and the intruders from nuclear physics mostly, but not entirely, returned to their labs. Some who did not became very influential in aviation's technical community. George Comstock, for example, who completed the GCA project after Alvarez and Johnston left for Los Alamos, was one of the few who stayed in aviation for the rest of their lives. He joined the Airborne Instruments Lab, became its president in 1960, and served on President Kennedy's Project Beacon commission, formed to recommend a new air traffic control system.

More important than the individuals that the lab gave to aviation, and far more controversial, was its bequest of the ground-control model. Before World War II, pilots' authority and autonomy in the air was unquestioned because no one but the pilots knew where they were and what the status of their aircraft was. The rudimentary air traffic control system that the Bureau of Air Commerce had begun piecing together in 1936 actually had no communication with aircraft at all, let alone any means to track them. It exerted no real control over pilots. Instead, it served as a means of exchanging information between airports as to expected arrivals and departures. It was, in its best days, a flight-following system, not really a control system. The wartime development of radar, however, promised—or threatened, depending on one's point of view—to place pilots under continual surveillance and control from the ground. The intrusion of Lawrence's cyclotroneers thus dramatically altered the technoscape of American aviation during World War II.

The war itself, however, was the driving force behind these changes. Without World War II, there would have been no Radiation Lab to challenge the dominant model of control. The war caused politicians to make available far greater resources for technological development than had previously been contemplated, and that in turn enabled investigation into a variety of different possible landing aids that would never have been pursued otherwise. The war established conditions within which unparalleled creativity was possible and rewarded, and in which any landing aid could get a trial by fire. The politics of blind landing, in essence, were suspended during the war because the need was so great that the AAF was willing to try virtually anything that promised to work better than what little it had, and it had the money to spend on multiple avenues of research. It also had no competition for control over aviation during the war and therefore did not have to negotiate with the airlines or with the CAA. The army could dictate the type of infrastructure created to serve the needs of wartime aviation without consideration of other organizations' needs, and equally importantly, AAF leaders came to rely upon the advice and expertise of academic scientists to help them make technological decisions. The war thus radically altered the environment in which landing aids research was done.

Wartime suspension of the politics of blind landing did not last longer than it took the ink to dry on the Japanese surrender document, however. A vicious political battle took place after 1945 over the selection of a common landing aid for the United States. Chapter 4 examined the negotiation of a standard system before the war to suggest how difficult reaching agreement could be even when all parties agreed to the basic form of the desired solution. After the war, no such agreement existed, with some organizations supporting GCA, and others supporting SCS-51. With the appropriate model of the solution itself open to question, the choice of a landing aid became deeply political.

CHAPTER SEVEN

The Politics of Blind Landing

In May 1947, Charles V. Murray told readers of *Life* magazine that the Civil Aeronautics Administration (CAA) had "declined into a mental outlook more befitting the guild of harness-makers." At issue was CAA's continued insistence on deploying its "old-fashioned" instrument landing system (ILS), which differed from the army's wartime SCS-51 only in that it was housed in buildings instead of in a truck, and not the army/navy AN/MPN-1 ground-controlled approach (GCA) system. GCA's glories had been trumpeted throughout the U.S. aviation press as the best solution—and by *Aviation News* and *AOPA Pilot* as the only solution—to that now-ancient bugaboo of fliers, the blind landing.[1] Criticism of CAA's "anti-radar" policy began in late 1945, almost immediately after cessation of hostilities, and became increasingly vitriolic throughout 1946. The August 1946 decision of the Collier Trophy Committee, custodians of the most prestigious aviation honor, to present the 1945 award for "the greatest achievement in aviation in America" to Luis Alvarez was merely one more blow to CAA's credibility. Over the next few months, Congress joined the media criticism and launched an investigation into air safety.[2]

Part of the controversy descended from the landing aids decisions made by the U.S. Army Air Forces (AAF) and the U.S. Navy. The navy had decided by late 1945 to abandon SCS-51/ILS completely in favor of AN/MPN-1/GCA, while the AAF publicly promoted GCA.[3] Nonetheless, the AAF continued to install both mobile SCS-51s and fixed ILS stations throughout the United States and along overseas routes. The majority of its GCA sets, moreover, stayed in storage despite its own rhetoric. Manpower shortages kept the army from utilizing the GCA sets it possessed, regardless of policy decisions. The AAF had chosen not to choose, in essence.

The aeronautical press bore a bias at least as deep as the one it attributed to CAA. It took at face value the words of GCA enthusiasts, even when their "facts" were somewhat questionable. The historian will search in vain through the pages of *Aviation News*, for example, for a single positive mention of ILS, despite its solid war record. Instead, one will find page after page devoted to reporting on the number of airmen "saved" by GCA. Only by referring to the National Archives can one discover that there were more than twice as many ILS systems in successful operation at military airfields in the United States in 1946 as GCA sets. Biased reporting created much heat but shed little light on significant issues.

The CAA's administrator in the first post–World War II years was Theodore P. Wright, a well-respected aeronautical engineer. Appointed in August 1944, he replaced Charles Stanton, who then became assistant administrator. Wright had come to Roosevelt's attention through his service as a member of the Aircraft Production Board, where he had earned a reputation for working well with the military. He also had foreign contacts and was a member of the Royal Aeronautical Society, which President Roosevelt had hoped would better enable him to work with the new Provisional International Civil Aviation Organization. Wright was also a believer in what Joseph Corn has called the "winged gospel," and he wrestled with the inconsistency between his interest in private aviation and his duty to the primary users of the airways, whom he felt were commercial and military fliers.[4] As a result, he spent most of his term being attacked by the Aircraft Owners and Pilots Association for not doing enough to help the "little man" gain access to the airways. The ILS/GCA debate is one place where Wright most spectacularly failed private fliers.

The issue of access to the skies is a deeply political one. Both GCA and ILS worked as effective landing aids, but *how* they worked—not how *well*—caused private and some military fliers to line up in the GCA camp, while the Air Line Pilots Association and most airlines supported the CAA and its ILS. In short, GCA was easier and cheaper for an "average" pilot to use, which is precisely why airline pilots were opposed to it. GCA enthusiasts saw it as a way to promote a democratic vision of aviation, one in which every adult could own and operate an airplane. The Air Line Pilots Association, by contrast, saw it as a threat to pilot autonomy and a means of deskilling the profession. Ultimately, however, the outcome of the controversy was determined by how the two systems allocated costs.

THE CONTROVERSY

The great debate over ILS and GCA took place in the aeronautical and mainstream media between December 1945 and the end of 1947, and in a congressional investigation into aviation safety which ran from January to March 1947.[5] The media universally took the pro-GCA position, granting no quarter to ILS supporters, all of whom were painted as "conservatives" refusing to adopt or even seriously consider new technologies, particularly radar. The leading GCA supporters were J. B. Hartranft Jr., head of the private-aviation-oriented Aircraft Owners and Pilots Association (AOPA) and editor of its newsletter, *AOPA Pilot;* and an aviation writer named William Kroger, who wrote primarily for the magazine *Aviation News*. The defenders of ILS were CAA's "old guard," led by Charles Stanton and the president of the Air Line Pilots Association, David Behncke.

A cartoon appearing in *AOPA Pilot* in June 1946 illustrates the participants and their positions in this rather complex process, at least from Hartranft's perspective (Figure 7.1). The man in the flight suit represents private fliers (and by implication, all fliers, as no ALPA character appears), while the "fat cat" businessman next to him represents the CAA. The absence of a figure for the Air Transport Association (ATA) implies that the CAA figure represented both the agency and its preferred client, ATA. CAA's officials were hardly well paid; Wright's salary went up when he left the administrator's office for Cornell University. Clearly, AOPA considered CAA biased toward business and combined CAA and ATA into a single figure.

The cartoon also accurately represents the positions of the armed services. A navy officer seeks to help convince the reluctant agency that GCA was the right choice. The navy had adopted GCA for several reasons. First, the ILS was unusable aboard aircraft carriers, as the beam required a stable reflecting surface. Second, the navy had always used a form of "ground-controlled approach" aboard aircraft carriers. A safety petty officer guided aircraft aboard using light wands and hand signals, so that Alvarez's GCA was merely an extension of its existing operating procedures. Third, as the navy's head of aircraft research told an interviewer, GCA was cheaper from the navy's point of view. Admiral Luis de Florez explained that while the GCA ground installation cost twice as much as ILS, the cost of the onboard ILS receivers greatly outweighed the ILS ground installation's lower cost. But the real savings from GCA came from reduced training costs. ILS required training thousands of pilots and, since skills atrophied when they went

Figure 7.1. "Only GCA can save us!" The Aircraft Owners and Pilots Association's view of the ILS/GCA conflict. *AOPA Pilot*, June 1946, 52a. Courtesy of Hagley Museum and Library.

unused, necessitated recurrent training every few months. GCA had no pilot training costs, only operator costs. Hundreds of pilots might use a GCA installation in the course of a day, but a new two-man GCA being developed would need only eight operators and one maintenance technician to serve them round the clock.[6] From the navy's point of view, GCA was an economical choice.

While the navy's position in the conflict was unambiguous, the cartoon makes clear that the Air Force's was not. The agency's leadership refused to take a public position on the issue, arguing instead that the two systems should be seen as complementary, not as rivals. This reflected internal arguments over the two systems' relative merits. In 1945, the AAF's chief instrument instructor pilot, Lt. Col. J. B. Duckworth, had argued that neither system was completely suitable and recommended further development. The agency then ran a series of tests during 1946 on the two systems at the Civil Aeronautics Administration's Indianapolis center which resulted in a similar set of recommendations. In its review of those tests, the Air University, which was charged with professional education of Air Force officers, argued that GCA, while it had put in a better showing during the test series, was problematic because it violated the principle of "air leadership" by placing control of aircraft in ground controllers' hands—precisely the pilots union's argument. It also recognized that ILS required greater levels of training, was more difficult to fly, and depended on special equipment in the aircraft. The Air Force's leadership ultimately decided that it wanted a fully automated system that allowed pilots to decide whether to use it or not; until it developed that system, it supported installing both ILS and GCA.[7]

Two major players are missing from the cartoon. The lack of representation for ALPA, the commercial pilots union, suggests that AOPA did not wish to reveal division among pilots over the two systems. It may also indicate that AOPA considered ALPA illegitimate. The two organizations rarely saw eye to eye, with the union consistently fighting for tighter regulation of the industry, while AOPA preferred a laissez-faire approach. AOPA may have felt that commercial pilots did not deserve separate representation. The other missing party, of course, is the U.S. Congress. That omission is forgivable. No one could have foreseen the political attention that the great ILS/GCA controversy eventually drew, but perhaps someone should have. Ultimately, the choice between the two systems was a political one, and thus well within that body's traditional rights.

If I have accurately represented the positions of the groups involved in the blind landing controversy, no single issue dominated the debate. Commercial pilots perceived in GCA a threat to their professional autonomy, and private pilots saw it as a means to safer (and less expensive) personal flying.[8] Autonomy was never an issue for navy pilots.[9] Cost was that agency's major concern. And autonomy was just one of several issues for the Air Force, while cost appears to have been irrelevant. Different issues mattered to different organizations.

The great controversy began inauspiciously, in routine meetings of the Interagency Air Traffic Control Board, which had been formed to standardize approach procedures for airfields throughout the country. In meeting no. 359, held on October 25, 1944, the members voted to allow the installation of what the AAF referred to as the "instrument low approach" system without restriction.[10] (The Army Air Forces had decided there was no such thing as an "instrument landing system," and this was simply the army name for CAA's fixed ILS installations.) Before that action, each installation had to be individually reviewed and approved, slowing installation work. Production problems that had slowed the CAA-AAF program to install fixed ILS stations during the war had been overcome, and the army had finally begun providing CAA with complete sets of ground equipment, including the AAF's straight-line glide path. Because many of the installations were being made at commercial airfields that had been taken over by the AAF at the onset of the war, CAA expected both fields and equipment to revert to commercial use very shortly after the end of the war. That would allow almost immediate availability of a standardized landing aid to its airline customers. The end of CAA's decade-long quest to improve airline regularity appeared, finally, to be in sight.

A few months later, in May 1945, CAA's bubble burst. The Interagency Air Traffic Control Board, which was dominated by the army and navy, reversed itself and proposed immediate cessation of all installations and commissionings of the instrument low approach systems, "until a standardization policy applicable to all systems could be established." On the record the reason given for this change in policy was to resolve some conflicts over frequency allocations and flight pattern interferences between adjacent stations. The internal CAA memo reports, however, that the real reason for the sudden reversal of the earlier decision was "that an important segment of military opinion favors discontinuance of the Instrument Approach Program by the Army and proceeding with installations of Ground Controlled Approach Systems of the radar type."[11] CAA's representative arrived at this conclusion after an informal conversation with Lt. Col. Clarence B. Sproul, the senior Army Air Force member. CAA's representatives strongly opposed the recommendation, of course. They pointed out that CAA's system was standardized, according to previous agreements established years earlier. It was congressionally approved and funded, which, they insisted, made carrying out the program mandatory. CAA was able to block final approval of the new policy, but its representatives recognized their organization suddenly faced a real problem.

The ground-controlled approach system was not unknown to CAA, although it was certainly unfamiliar to many of its personnel. Thomas Bourne, the director of Federal Airways, had accidentally been invited to the official tests that the MIT Radiation Lab had held at National Airport in February 1943. He had been required to sign papers barring him from discussing it, even within CAA, but by 1944, GCA was an open secret within the organization. Except for that early test series, however, no one else in the organization had seen it or operated it, let alone learned its details of operation and maintenance procedures. Regardless of the system's performance, CAA was not prepared to operate it in February 1945.

During the controversy, accusations flew regarding who knew what, and when, with ground controlled approach partisans attempting to depict CAA as incompetently slow even to try the system. Thomas Bourne was attacked because he had known about GCA since early 1943 but had not done anything about it, despite the wraps of secrecy around the system through late 1944.[12] Bourne did not help matters any by claiming that he was unaware of GCA before February 1945, which was actually the date CAA was first allowed to send staff to train on and evaluate a GCA installation. *Aviation News* also reported Maj. Gen. Harold McClelland's claim that he had offered Bourne a GCA set during the fall of 1944 but had been refused, to further paint Bourne into the "old guard" anti-GCA camp. The record, however, shows that McClelland had offered to let CAA evaluate a GCA set in November and that Bourne had accepted it the following month. Those tests were scheduled for February at Bryan; that same February Bourne also formally requested that the Army Air Forces loan a GCA set to CAA for more extensive tests and to serve as the basis for drafting specifications for a civil version.[13] Bourne also met with the president of Gilfillan Brothers, the army's ground controlled approach manufacturer, and Homer Tasker, Gilfillan's GCA project engineer, in February to arrange technical training and maintenance assistance. The requested unit arrived in May at the Indianapolis Experiment Station, still nominally an AAF-run facility. Kroger, the leading GCA partisan among the aviation writers of the day, later summarized this exchange in an "objective" history of the controversy, which consistently portrayed McClelland's claims as the true version of events.[14]

The record seems to show that CAA moved as quickly as one can expect of a large bureaucracy, especially given its almost complete isolation from the vast array of new aviation technologies developed during the war. As agency historian John Wilson has pointed out, CAA had started the war ten years ahead of the AAF in navigation technologies and ended it ten years behind. That was hardly the agency's fault. All of the new technologies were secret and, once unveiled, took time to learn. They also required money, and CAA's budget for radio and radar research was a mere $87,000 for FY1946, and $180,000 for FY1947, which Ben Stern, the assistant administrator for public affairs, told the editor of the *Chicago Times* was one-fifth what the agency had requested.[15] The amount was not enough to buy a set of any type of radar, and CAA was therefore entirely dependent on handouts from the Pentagon for its investigations. It obtained its first radar gear from the navy, which sent ten truckloads of various sorts of equipment in late 1944 to the Indianapolis Experiment Station. That delivery is what had prompted Bourne to try to obtain a GCA set, which the navy had not included in its bequest.

Glen Gilbert, chief of CAA's Air Traffic Control Division, conducted the agency's first evaluation of GCA during February 1945 at the Army Air Force's Instrument Instructor School at Bryan, Texas. Gilbert had been one of the first group of controllers hired by CAA to run its first experimental "en route" air traffic control center, which went into operation in 1936.[16] His analysis was certainly biased by his job assignment, and he reported primarily on GCA's utility as a traffic control aid.

Gilbert described discussions he had held with various staff members at the Instrument Instructor School. According to his report, they believed that because the final controller could handle only one aircraft at a time, the maximum landing rate for GCA was one plane every two minutes, based on the two-mile range of the precision radars. Additional precision scopes and an extra controller could raise this capacity to one per minute, which "was just about the capacity of a single runway under any weather conditions." By contrast, the instructors thought that SCS-51 could only handle one aircraft every three minutes, due to beam deflections caused by other aircraft. Although this appeared to give advantage to GCA, both CAA and AAF found that the two-mile approach the AN/MPN-1 offered was too short for the largest aircraft and increased the precision radars' range to six miles, and then to ten miles, reversing the capacity advantage.[17] In actuality, the capacity issue was never subject to public controversy, and throughout 1945 and 1946 both CAA and AAF sought to find ways to increase the safe capacity of both systems.

More important was Gilbert's evaluation of the ease of use of the two systems:

"Based on discussions with officers who have had considerable experience in both systems, and considering my personal reaction in making consecutive approaches under the hood (SCS-51 in an AT-6 and GCA in a B-24), it is the writer's personal opinion that making instrument approaches under GCA will be easier for the average pilot than when using SCS-51."[18]

Gilbert was seconded by A. H. Hadfield, the assistant chief of the Airways Engineering Division.[19] This had, of course, been one of the system's major selling points for the generals and admirals who had witnessed the official Radiation Lab demonstrations in February 1943. Alvarez had intended it to be easy for pilots, and Gilbert's statement reiterates that original goal. It was also a prophetic statement, and the first warning to CAA's old guard that they might face a public revolt.

By the end of the year, the two men found their opinions being echoed by members of the aeronautical press. William Kroger reported in December 1945 that "private flyers and segments of the industry [were] arranged against CAA, criticizing its instrument approach system as too complicated." He blamed CAA's conservative faction for continued promotion of ILS, and in subsequent articles various other *Aviation News* authors named Stanton and Bourne as the leaders of that faction. The Aircraft Owners and Pilots Association was the segment of the industry in question, as the Air Transport Association supported CAA's ILS. The two exceptions were Pan Am (not represented by the Air Transport Association, since it served overseas routes exclusively), and Jack Frye's TWA. Pan Am, along with a number of transatlantic airlines, rented a GCA set from the army and operated it at Gander, Newfoundland, beginning in January 1947. TWA did the same at Reading, Pennsylvania, as an emergency field for its major hub at Philadelphia. For that, the two received great accolades from *Aviation News, Popular Science,* and even the *New York Times.*[20] Adm. Emory S. Land, president of the Air Transport Association, pushed CAA to make experimental GCA installations, yet Land and his technical vice-president, retired general Milton Arnold, emphasized GCA use as traffic control aid, not primarily as a landing aid. Hence, "segments of the industry" is at best a misleading phrase, and many aviation writers seem to have followed Kroger in misconstruing support for GCA as a control aid as support for it as a landing aid.

Although the aeronautical press castigated CAA for its "anti-radar" faction, it was clearly an unfair characterization. Administrator Wright wrote to James Johnson, president of the Springfield (Missouri) Flying Service, that the CAA program included "the adoption of ground radar search installations for use in connection with airport traffic control."[21] He did not, however, mention its use as a landing aid. That distinction was never made clear within the aeronautical press, and a large portion of the acrimony over radar was probably derived from sheer confusion. CAA's policy was not anti-radar, but its officials by and large were opposed to its use as a landing aid.

Leaving aside the innate bias Bourne, Stanton, and others had toward ILS, the use of radar as a landing aid appeared to be a very bad deal to CAA. One problem it perceived was liability. If a pilot using ILS made a mistake and crashed, it was legally "pilot error" and therefore an airline responsibility. If the ILS were broken, aircraft could be sent elsewhere. If a GCA final controller made a mistake, then CAA clearly was responsible. Since pilots had no other source of information on their approach position in a GCA-only approach, they could not evaluate the controllers' orders, and hence CAA could not pin the blame on them.

A larger problem for CAA was the sheer number of operators that a full-scale GCA deployment would have entailed. The wartime sets, as described in Chapter 6, required five operators per shift, with at least one additional maintenance person. CAA worked with Gilfillan Brothers during 1946 to reduce that to two operators per shift, but for twenty-four-hour operation that still required eight operators and two maintenance crew per installation. The agency planned to install ILS at 180 airports throughout the country by 1950, which would mean an additional 1,800 new employees if GCA replaced ILS in its plans. That was a 10 percent increase over CAA's 1946 payroll. CAA was under no illusion that Congress would happily provide it with so many new payroll lines; indeed, the agency found Congress unwilling to provide sufficient payroll funds to bring it to its authorized manpower. Worse, Congress effectively cut CAA's payroll in 1946, prompting the closure of fifty-five airways communications stations.[22] Given its tight budget, it is unsurprising that CAA's policy was to use radar sparingly. Since ILS was automated, and required only one maintenance person whose duties could be split between the ILS and other airport electronics gear, it made far more sense to officials.

Then there was the issue of maintenance costs. Although CAA had received and tested a GCA set during 1945, it did not undertake anything like the rigorous testing necessary to really evaluate the system's performance or determine its true operating costs in civil use until 1946. Military cost estimates were essentially useless because military operations did not sufficiently resemble civil ones. With the exception of Military Air Transport Service operations, military aircraft took off and returned nearly simultaneously. That meant the system had to handle high volume, but only for short periods of time. The system could then be put in standby for several hours before the next wave of aircraft operations. Military operations did not require continuous availability, and maintenance could be scheduled between operating periods. Commercial use, by contrast, required continuous operation of the system for at least twelve and even, at the largest airfields, twenty-four hours a day. Maintenance time came only at the expense of service to aircraft, which meant commercial use required a vastly more reliable piece of equipment than the military needed.

CAA found during its investigations in 1946 that the Gilfillan-built AN/MPN-1 it had been loaned required 900 hours of maintenance for 1,600 hours of operation, although some of that maintenance could be conducted while the system was in operation. Gilfillan Brothers hotly contested that figure, but based on the Signal Corps' own maintenance records for the twenty-eight GCA sets it operated during 1945 and 1946, it may have been accurate. There was extreme variability in the number of maintenance hours needed to keep the equipment running: some units spent more time being maintained than being operated, while others appear to have been very reliable.[23] One likely explanation is poor quality control at the factory. With 600 vacuum tubes stuffed into a trailer, minor variations in placement, quality of soldered connections, and quality of components could cause large differences in reliability between units. CAA might have gotten a lemon. Another explanation is quality of maintenance personnel. Inexperienced or poorly trained maintenance people require much more time to accomplish a given task than someone who knows the equipment well, and CAA's own lack of radar acumen no doubt drove up its tally of maintenance hours. Finally, the basic design of Gilfillan's equipment seems to have affected its reliability. During the war, Bendix engineers sent to Gilfillan Brothers by the navy to learn how to build GCA had substantial disagreements over design with Gilfillan's chief engineer; after the war, CAA found that Bendix-made GCA sets, known officially as AN/MPN-1(B), were easier to maintain and much more reliable.

Regardless of cause, CAA's officials were convinced that the maintenance requirements of the wartime GCA sets were too great for civil use. They worked with Gilfillan to try to reduce those costs. The agency reported its efforts in *CAA Journal* throughout 1946 and 1947. Re-engineering GCA for commercial service meant, of course, several years of delay before any feasible implementation of either full GCA or even just the search radar. Engineering work required time and money, neither of which were granted to the besieged CAA.

Cost, however, depends on one's point of view. As noted above, from the navy's perspective GCA was less expensive. The only highly trained people required by the system were the ground controllers, and even a large-scale GCA installation would require fewer operators than there were pilots who, in an ILS world, would have to be extensively trained, and constantly retrained, in order to maintain their skills. "Under such circumstances," Adm. Luis de Florez had stated, "GCA is incomparably cheaper . . . The training you give the ground operators of GCA covers literally the hundreds of pilots a day he [sic] can talk in safely."[24]

The admiral was most certainly correct, from a certain point of view. Eliminating a great mass of pilot training to train a much smaller group of controllers meant significant savings to the navy. For the CAA, however, which did not have to pay pilot training, the addition of controllers and increased training was a major budget blow. ILS minimized cost to the agency, at the price of raising the airlines' costs. The airlines did not object, since they expected additional revenue from whatever landing aid was selected, and most were not thrilled with various cost-sharing proposals that were bandied about in Congress to fund GCA. Hence the different economics of commercial and military aviation influenced technological choice in a way that was never made clear in the media. GCA probably was cheaper from a global economic standpoint, but such a consideration was irrelevant to the CAA, the Budget Bureau, and to the Congress.

Therefore, CAA did indeed object to radar landing but not solely because of the conservatism of Stanton's old guard. The agency believed it had legitimate concerns based on its funding history balanced against the costs it expected radar landing to impose over and above a much more limited traffic control radar deployment. The media's own bias, however, prevented it from explaining the real costs of radar landing. In fact, the media often misrepresented the issue. *AOPA Pilot*, for example, told its readers that one GCA set could serve Washington National, Bolling Field, and Anacostia Naval Air Station.[25] That, of course, was an absurd claim, but one the average person could not evaluate. Nor could the media advocates accept that re-engineering the wartime GCA was necessary to make it commercially useful, and they insisted that CAA adopt the trailer-mounted sets immediately.

Pressure thus built on Administrator Wright to do something about GCA during 1946, and he decided to try to obtain some additional GCA units to install at commercial fields in order to determine the true costs of operating them as commercial installations. He was probably persuaded to try this by Milton Arnold, an ex-brigadier general who had commanded a bomber group in Europe during the war and had left the army to become head of the Air Transport Association's engineering section. Arnold had broached the issue of the Army Air Forces loaning three sets of AN/MPN-1 equipment to CAA as well, and the AAF's internal response is worth repeating verbatim: "The primary purpose is to demonstrate con-

clusively to CAA the advantages of search radar in simplifying airport traffic control and at the same time improve airline schedules and operations during the
coming winter. Early adoption of airport radar by the CAA will be advantageous
to the AAF as well as the airlines, consequently this Headquarters [Air Materiel
Command] concurs in the desirability of aiding in the project even to the extent
of loaning developmental equipment."[26]

Air Materiel Command's letter suggests that neither the Air Transport Association nor the Army Air Forces was dedicated to forcing CAA to adopt GCA's
landing function. It clearly is aimed at GCA's traffic control function, in the same
way that Glen Gilbert of CAA's Air Traffic Control Division had emphasized the
search radar's utility in his 1945 report. That traffic control was the important issue to the Air Transport Association's leaders is unsurprising, despite the media's
devotion to radar landing. Its officials were well aware that if CAA suddenly
dropped ILS for radar landing, widespread deployment would be several years
away due to the vagaries of federal budgeting and the need to re-engineer the
equipment. ILS was already installed at thirty-nine major commercial fields, and
CAA had funding for a large number of smaller fields as well. The only reason it
was not already operational was lack of receivers for the aircraft, and those too
were expected in a few months. The ATA and its member airlines had been
burned before by waiting for something "better" to be perfected when a "good
enough" system already existed, and it was not about to make that mistake again.
From the airlines' point of view, radar landing was not worth waiting for.

However, the letter also implies that Air Materiel Command agreed with the
aeronautical press that some elements within CAA were not giving GCA's search
function its due. Both Wright and Glen Gilbert, however, were already enthusi-
astic about GCA's possibilities as a traffic control device. Agency poverty prevented them from doing much about it, although Wright did seek to have CAA's
radar research budget increased. The Budget Bureau also denied this request. So
borrowing was all he really could do.

Wright happily accepted the Army Air Forces offer to loan three GCA sets for
installation at Chicago, La Guardia, and Washington National after getting the Air
Transport Association to agree to pay the installation cost (a not-inexpensive
$20,000 per site) and arranging for Gilfillan Brothers to install three experi-
mental remoting kits to place the radar screens in the airport control towers instead of in the trailer parked on the field. Those kits were also part of the AAF's
loan. Where the money for their installation came from is not clear in the records,
but CAA had to pay the twenty-one operators necessary to run the three stations,
borrowing people from other parts of its airways operation service to crew the

units.[27] The GCA sets did not come back from Gilfillan until late in the year and were not ready for service before January 1947.

Despite Wright's efforts to overcome whatever bias existed within Stanton's old guard, it was never enough to satisfy the GCA enthusiasm Kroger and other aviation writers promoted. He could not have satisfied it because he worked within constraints imposed on him by CAA's position as a poorly funded subunit of the Commerce Department. His primary constituency, the commercial airlines, wanted a system of landing aids immediately, not years in the future, and ILS was all that was available. Radar landing, even if he had been able to reconcile its high cost with his comparatively small budget through some means of creative cost-sharing with the airlines, was years away from service if for no other reason than the time needed to re-engineer the equipment to reduce its maintenance and operator costs. During the war emergency, when cost was no object, GCA had taken a year and a half to get into production, and the aviation writers' presumption that a civilianized, rigorously tested radar system could be had in mid-1946 was absurd. Ultimately, the GCA enthusiasts could not see that "combat proven" was not a good enough appellation for civil use. Civil aviation had different needs.

Yet CAA's concerns about radar landing were not entirely legitimate, either. The budget problem CAA's officials perceived was not really CAA's problem at all. It belonged to Congress, which had the authority to fund whatever it chose to approve. CAA's budgetary concerns, although perhaps honorable, were thus somewhat misplaced. As one member of the House Interstate Commerce Committee soon pointed out, Congress was ultimately responsible to the public for the air transportation system, not the unelected CAA. Part of the agency's annual responsibilities was to provide a report on what it needed to safely maintain the nation's airways so that Congress could then make informed decisions in fulfillment of its duty. In this view, CAA had prevented Congress from exercising its responsibility to the flying public by proposing only ILS and trying to ignore radar landing. The agency's devotion to ILS thus seemed to many members of Congress to be protection of a pet program, not pursuit of the best technologies. By late 1946, the media had convinced Congress of GCA's superiority, and Congress joined the anti-CAA alliance.

CONGRESS ENTERS THE FRAY

The midterm election of 1946 returned the first Republican majority to Congress since Herbert Hoover left office in 1932 and thus provided the disenfranchised Republicans their first opportunity to roll back the "creeping socialism" of the New Deal and attack Truman and the leftover New Dealers in his administration. The Republican leadership probably perceived that Harry Truman was particularly vulnerable to an investigation into his administration's handling of aviation matters, since as a U.S. senator he had been a leading advocate of the Civil Aeronautics Act that had founded CAA. He had also chaired a long-running investigation into defense procurement during World War II, which focused on the aircraft industry.[28]

The Republicans had two pretexts for mounting the investigation. The major newspapers and the aeronautical press had noticed a "rash" of fatal air crashes during 1946, which the airlines claimed had resulted in a sudden drop in air travel. A number of articles claimed radar or GCA could have prevented many of these. The acrimonious debate over the two landing systems, which of course involved the much-criticized lack of radar along the airways, provided the other excuse for a full-blown congressional investigation.

The rash of accidents, in reality, was a statistical artifact. The commercial airline accident rate for 1946 was actually only about half that of 1945, but because the amount of traffic had nearly tripled, the number of accidents went up, and accordingly, so did media coverage (see Table 7.1). The crashes, according to United and Eastern Airlines spokesmen, had resulted in a "marked decrease" in passenger traffic from the previous year. Either the airline spokesmen were unaware of the statistics that their superiors gave CAA or they chose to project an image of reduced safety in order to "help" CAA get more money from Congress. Regardless, the airlines' claims inspired great leaps of congressional rhetoric. Rep. L. Mendel Rivers (D-SC) declared, "The American people are horrified and scared to death."[29]

The perception that CAA was not doing its job was further reinforced by claims that radar (in the *New York Times*) or GCA (in *Aviation News*) could have prevented some of the crashes.[30] The airlines had contributed to the perception

TABLE 7.1
Fatalities per 100 million passenger miles

Year	NATS	ATC	US Airlines
1944	18.1	5.5	2.1
1945	19.7	6.0	2.1
1946	10.8	8.2	1.2

Source: CAA Statistical Handbook, 1950.
   Notes: NATS is the Naval Air Transport Service, while ATC was the Air Transport Command.

of danger by trumpeting the merits of wartime radio and radar devices as safety advances, without making clear that these were years from being ready for actual commercial service. The public expected the number of accidents to drop as a result of these advances, but instead the number went up. Nevertheless, the aviation system was safer. It simply did not appear that way to a public misled by a number of sensational crash stories and equally ostentatious tales of radar's "all-seeing eye." Congress chose to act on the appearance rather than the reality.

The second pretext for Congress's investigation into CAA's management of the aviation system was the ILS/GCA controversy, which had been rekindled in late November 1946 when the *New York Herald Tribune* and *Chicago Tribune*, as well as *Aviation News*, published parts of some allegedly "secret Army Air Forces' tests."[31] These reports cited Lt. Col. Clarence B. Sproul's claims for GCA's superiority over ILS and provided much scientific-appearing data to back him. The appearance of these pro-GCA articles marked the beginning of a full-blown public controversy. Combined with the rash of air crashes, they virtually ensured congressional attention to the landing aids issue.

The House Committee on Interstate and Foreign Commerce, which oversaw the Commerce Department, convened hearings in January 1947 to investigate the air safety crisis and CAA's management of the federal airways. The full committee was chaired by Rep. Charles Wolverton (R-NJ), while Carl Hinshaw (R-CA) chaired the subcommittee established to run the air safety hearings. In general, the transcript provides little evidence of political grandstanding, and the members seem to have been genuinely concerned with the issues in question, although the majority clearly did not understand the technologies involved. Hinshaw, whose Pasadena-based district contained substantial portions of the aviation industry and had long been interested in aviation matters, was one exception. Also, M. Harris Ellsworth (R-OR) clearly recognized the political nature of the controversy.

The air safety subcommittee heard a veritable parade of witnesses from the aviation industry over the next two months. Generals Ira Eaker and Harold Mc-Clellan were among the earliest witnesses, while the navy's primary witness was Rear Adm. J. W. Reeves, head of the Naval Air Transport Service (NATS). Wright, and his superior in the Commerce Department, William Burden, were CAA's major representatives, while Milton Arnold represented Air Transport Association and David Behncke spoke for the commercial pilot's union, ALPA. Last to speak to the committee was AOPA's Hartranft, who was asked no questions.

While the army's General McClelland carefully avoided taking a position on the great controversy despite several attempts to pin him down, Admiral Reeves was happy to discuss the navy's decision to use GCA exclusively. It had, he reported, enabled NATS to reduce its weather delays to an average of 2 percent of all flights. He also told that committee that in January 1947, in a period during which 45.7 percent of total commercial scheduled operations in the San Francisco area were canceled, only 1 percent of NATS operations at GCA-equipped Moffett Field were canceled.[32] Reeves also referred the committee to Colonel Sproul's report, which was entered into the record as evidence. Publication of Sproul's test reports in the hearing transcript marked the first public appearance of the full report and no doubt helped convince the committee of CAA's backwardness on the issue.

Clarence Sproul, now an ex-lieutenant colonel and the same man who had informed CAA's Interagency Air Traffic Control Board (IATCB) members of the AAF's sudden change of policy toward GCA in May 1945, had characterized his tests for the media as "proving" the superiority of GCA's radar landing function over ILS. Pilots using GCA, Sproul reported, "achieved 350 per cent better runway alignment than those following the glide path-runway localizer beams of the I.L.S." The most detailed publication of Sproul's findings was in *Aviation News*. William Kroger reported that airline captains who participated had an average of 6,560 hours flying time, 40 percent had previously flown GCA, while 20 percent had flown ILS. Ninety-one percent were able to touch down with GCA, while only 54 percent succeeded with ILS. A group of fighter pilots, Kroger reported, flying the C-54 test aircraft for the first time, and with an average of only 805 flying hours, "were far more accurate in landings on GCA than with ILS."[33]

Before trying to glean anything meaningful out of Sproul's data, some discussion of the test methodology is necessary. Unsurprisingly, neither of the writers bothered to ask about how the tests were conducted, since "scientific tests" were supposed to be objective and report the truth. Sproul's tests, however, contain a serious methodological flaw that renders the details suspect. The AN/MPN-1 set's precision radars were used to evaluate the performance of both themselves and the ILS.[34] There was therefore no independent verification of either system and thus no adequate control despite the selection of "control groups" of pilots. If GCA happened to be in error, for whatever reason, the error could not be detected.

In response to publication of Sproul's tests, Wright informed the aviation editor of the *Boston Traveler* that CAA had yet to find "a single case where any pilot using the instrument landing system missed the runway at its end by *as much as* 326 feet, which Mr. Sproul reports as the *average* error for pilots experienced in ILS."[35]

CAA's director of Air Navigation Facilities Service contended that an error that great at touchdown represented a 3.3 degree deviation from the landing course, which would result in a full-scale deflection of the cross-pointer instrument. No certificated pilot, he claimed, could be so consistently poor.[36] Yet airline pilots did even worse than the control group average, according to Sproul's data, achieving a dismal average error of 726 feet.

Despite the bias of CAA's people toward their own gear, they were certainly correct in their belief that no pilot who had survived 6,000 or more flight hours could be so bad. The poor showing of ILS was probably due to a misalignment between the two systems, which would cause GCA's performance to appear far better than SCS-51's. Sproul certainly had no means to ensure such precise alignment, and in actual operation, very precise alignment is unnecessary if landing aircraft is to be done visually. Only automatic systems require the kind of precision alignment necessary to eliminate the possibility of beam divergence. Hence Sproul's detailed data is at best unreliable, and it certainly misled both the media and congressional investigators as to the relative performance of the systems.

Nevertheless, some of Sproul's data remains useful. One piece of data in each series, the "percentage of Approaches from which a Landing Could Have Been Made" is based in part on input from the check pilots, who were able to see the airfield. That eliminates some, but not all, of the pro-GCA bias in the numbers. Since only a few diehards within CAA still considered blind landings feasible— Sproul's report rejects it explicitly—pilots would make visual landings after breaking through an overcast. Once CAA placed ILS in commercial service, it imposed a 300-foot ceiling, reducing it to 200 feet for each airline as its pilots gained experience. Hence, the 200-foot data permits some useful conclusions (Table 7.2).

TABLE 7.2

Percent approaches from which a landing could have been made

Test Group	GCA	SCS-51
Control Group #1	100%	83%
Control Group #2	100	92
Air Transport Command	97	88
Air Material Command	94	61
HQ, Army Air Forces	93	64
Bolling Field	91	75
Fifteenth Air Force	87	79
CAA	100	91
Air Line Captains	100	72
Training Command	100	100
All Pilots	95	80

Source: House Committee on Interstate and Foreign Commerce, *Safety in Air Navigation*, 80th Cong., 1st sess., Jan. 1947, 531–37.

All pilot groups achieved better results on GCA than ILS, with the exception of the AAF's Training Command instructors, who achieved 100 percent on both systems. The worst GCA performance was 87 percent, by the fighter pilots of Fifteenth Air Force, who were flying a C-54 transport, an unfamiliar aircraft of much greater weight than the fighters to which they were accustomed. That certainly made flying both systems more difficult, as they would not have been able to accurately judge the aircraft's handling. All other pilot groups achieved above 90 percent. GCA direction was thus very consistent in its performance regardless of the experience of the pilots flying it.

The SCS-51 results, in contrast, vary from a high of 100 percent for the Training Command pilots to a low of 61 percent for the Air Materiel Command pilots. The worst-scoring AAF pilots were the least experienced in instrument flying. Especially telling is the 64 percent showing for the AAF headquarters pilots. Although they had a relatively high flight hours average, they were in nonflying billets and had not flown much in the ninety days preceding the tests. The poor showing indicates that their instrument skills had atrophied. Their much better performance under GCA in turn suggests that Glen Gilbert's 1945 assessment of GCA's greater ease of use was correct. The SCS-51 required more training to use successfully, and it required that pilots use it more frequently in order to maintain their skill level.

GCA was precisely what AOPA and its general manager, J. B. Hartranft Jr, wanted in a landing aid. Hartranft, a former AAF officer, had been the other AAF member of the Interdepartmental Air Traffic Control Board when Colonel Sproul had pushed for the cessation of ILS installations in early 1945. He declaimed against ILS and CAA's old guard from the pages of *AOPA Pilot,* emphasizing that GCA would cost fliers nothing if adopted. With all equipment on the ground, provided by CAA, private fliers would have nothing to buy and no training to pay for.[37] To the House Committee, he emphasized that the receivers were too large and heavy for private aircraft to carry. They were also too expensive at an estimated $750 apiece, in an era when an airplane cost about $1,500. They were not even available yet, and would not be for private fliers for several years. ILS was useless to the members of his organization.

Hartranft then argued that ILS would not serve the majority of airways users. According to his numbers there were 80,000 private aircraft in the United States, and only 800 aircraft operated by scheduled air carriers. The airlines thus did not constitute the vast majority of the flying public. Since ILS could only be used by the 800 commercial aircraft, taxpayer money should not be spent on it, especially since GCA was the "proven superior system."[38] He therefore demanded that Congress act to stop CAA's ILS installation program, force it to dismantle the installations already in existence, and buy GCA.

Certainly GCA was the better choice for private fliers, for both economic and safety reasons. Implicit to Hartranft's argument, however, was an assumption about who, exactly, counted as an airways user. He had pointed out that "the amount of cross-country navigational flying accomplished by personal aircraft owners and private pilots is much in excess of that accomplished by commercial carriers."[39] Pilots, he assumed, were the only airways user, and therefore private pilots were the dominant user, with military pilots second. The mere 7,000 commercial pilots were a distant third. The air navigation system should thus be designed to suit private pilots, and he railed against CAA's "reverse thinking" in favoring commercial airlines and for choosing to solve only the problems of commercial users.

Wright refuted Hartranft's claim that private fliers were the majority of airways users, explaining that the "person-hour" was the best measure of airways use and that "the individual passenger in a civil or military transport aircraft is just as much a user of the [air navigation] system as the pilot of a military aircraft or personal aircraft."[40] Therefore, of the 173,000,000 person-hours flown in 1946, 163,000,000 were flown in commercial and military aircraft, which could be equipped for ILS (see Table 7.3). In Wright's formulation, the majority of users were passengers on commercial aircraft. The air navigation system should therefore be built primarily to serve that user group most efficiently, despite the imposition of higher costs on the private fliers.

AOPA and CAA could not even agree on who airways users were. AOPA's bias, of course, is obvious. But CAA was supposed to represent all users, and its chosen statistic, the person-hour, did show a clear commercial bias. A large commercial aircraft, with many more seats, longer range, and higher speed, could rack up "person hours" a great deal faster than any private aircraft could hope to.

TABLE 7.3
CAA Airways use statistics

Group	Millions of Person-Hours Flown in 1946
Scheduled Air Carrier	99
Non-Scheduled Air Carrier	27
Army	20
Navy	17
Personal	10

Source: House Committee on Interstate and Foreign Commerce, *Safety in Air Navigation,* 80th Cong., 1st sess., January 1947, p. 145.

The person hour was not the only statistic CAA had to choose from. Admiral Reeves, for example, had taken other CAA numbers and cleverly demonstrated that his Naval Air Transport Service was the biggest airways user, before explaining to the committee that anyone could make the numbers prove any particular case.[41] Wright could have chosen some other statistic and thereby made a different argument. The case Wright chose to prove, however, was the commercial one. More than any other act Wright could have performed, his choice of the person hour reveals his agency's bias.

Wright's defense of commercial aviation certainly satisfied ILS's most outspoken defender, the Air Line Pilots Association. ALPA had been founded secretly in 1931 by David Behncke, who was still its president, after the failure of an earlier attempt to unionize commercial pilots.[42] Behncke was a labor activist, but he refused to involve ALPA with other labor organizations. He wanted pilots to be seen as skilled professionals, not as common laborers. Behncke's protection of the image of pilots as skilled professionals helped the union immeasurably in its aviation safety campaign, which Behncke waged personally in ALPA's monthly newsletter, through friends in Congress, and occasionally in more mainstream publications.

Behncke chose to attack GCA on the issue of safety. In the March 1946 issue of *The Air Line Pilot*, he had written to his fellow pilots: "'GCA' requires complete and precise coordination of five operators on the ground and the pilot in the airplane in order to effect a safe landing. Think this over for a minute and realize the possibility of human error involved plus the possibility of malfunctioning equipment."[43] Behncke believed that the additional humans in what we would call the "information loop" was bound to increase the probability of human error and thus cause more accidents than the mostly automated ILS. Because of this, he argued, ILS was the safer system. He moderated somewhat in his congressional testimony (where he advocated proper runway lighting as the most important landing aid), and in a later article in *Air Transport*, the official organ of the Air Transport Association (and therefore a management publication). There, he apparently decided to cleave more closely to the airlines' official position that GCA should be adopted for traffic control and ILS should be the landing aid.[44]

In his own newsletter, however, he published the results of a poll he had conducted of the union's fifty-three regional councils. Forty-six of the councils had voted for ILS as the primary landing aid, while only one had supported GCA. The remaining six had abstained, because their members felt they did not have sufficient experience to properly evaluate the systems. Behncke also noted that seven councils had written to support the use of GCA as a traffic control aid, although they had not been asked their opinion on that subject.[45] Behncke thus had overwhelming support from his members for his ILS campaign.

ALPA's public posture promoted ILS as the safer system, but Behncke also had other motivations. Historian Nick Komons has argued that ALPA is a schizophrenic organization, which uses a public rhetoric supporting air safety to mask its real concern, the economic well-being of its members. ALPA is, after all, a labor union, and unions exist primarily to advance their members' economic interests. Yet unions also have tended to try to protect their members' skills from being eliminated by technological change. ALPA's defense of ILS was also a defense of pilots' skills. As Sproul's results suggest, ILS was indeed harder to use and required more training and practice than did GCA. It therefore maintained, if not actually increased, the skills needed to operate a plane, while GCA transferred some of those skills to a set of ground operators. Gilfillan Brothers and the AAF were already working on an automatic GCA, which was first demonstrated in 1947.[46] The union's pilots could thus look forward to a complete loss of landing skills within a few years if GCA were adopted. Automation of ILS, because the automation system resided entirely within the plane and was therefore under the pilot's control, was a far more acceptable outcome.

AOPA and ALPA, therefore, had fundamentally different interpretations of Luis Alvarez's GCA. One group perceived it as a means of improving their access to the airways system, while the other considered it a threat to their profession. The key issue Congress faced, then, was choosing which of these constituencies to favor. One member of the House committee recognized that this was, ultimately, a political issue. Rep. Harris Ellsworth tried to make that point clear while questioning Administrator Wright: "My point is the old question of how to spend money to do the most good for the most people. We are spending $6,000,000 [on ILS] for the benefit of 900 airplanes in the air today, whereas there are 50,000 planes which could use this other system of approach [GCA] . . . Why not ask more money and put in this other system which seems to be admittedly good?"[47]

Wright's only defense was to repeat that CAA believed ILS was the better system, especially if it could be supplemented by GCA at the busiest airports for traffic control purposes. The issue of better for *whom* did not, apparently, enter Wright's thinking, while Ellsworth clearly understood that was the root of the problem. Ellsworth rightly believed that GCA was better for the majority of pilots in the United States, while Wright followed the bias of his organization and its favored constituents, the commercial airlines, toward believing ILS was a better choice. Ultimately, Ellsworth sought to persuade his colleagues that they faced a political choice between different interest groups, not a technological choice over competing claims of technical superiority. The two models of operation, ground-controlled versus pilot-controlled, proved to be inherently political.

The committee chose to ignore Ellsworth, however, and focused on a different issue in its final report. The debate in the media had convinced the members that CAA's plans differed substantially from the Pentagon's. The members of the House Commerce Committee were very attentive to the needs of the AAF and navy in their 1947 investigation. In part, the members' bias toward national defense issues derived from the almost complete unpreparedness of the AAF for war in 1941 and an understandable reluctance to see that situation reoccur. And although there is no mention of a Soviet threat in the 1947 transcripts, the near-disintegration of U.S. armed forces during 1946 had not been matched by similar disarmament in the only other potential world power. The members were at the very least convinced that civil aviation had to be built to be easily available for military use in a future emergency, regardless of the source of that emergency, and they focused on CAA's well-publicized "disagreement" with the military over landing aids.

There was, in fact, very little disagreement between the Army Air Forces and CAA in terms of their landing aids policy. Indeed, Major General McClelland told the committee that the controversy was "unfortunate," and had diverted attention from their joint integration program, while former brigadier general Milton Arnold, of the Air Transport Association, said he thought that CAA's GCA work could not have been done better.[48] Mr. Ellsworth ruefully admitted that "there is a tendency on the part of some of us who are laymen on this subject to grab at these two combinations of initials and consider that there is a battle on as between the two of them."[49] But because the majority of witnesses the committee had heard took sides, and the press clearly had as well, the air safety subcommittee was already biased toward an either/or choice. Ultimately, it did consider the two systems competitors, basing its final report on that assumption.

The final report therefore criticized CAA for its consistent refusal to accept the decisions of the military-dominated Air Coordinating Committee (ACC), which had called for cessation of the ILS program in 1945. The committee stated that "any area of disagreement between civil and military aviation as to navigational and airport approach aids should be immediately resolved in the interest of national defense." The Air Coordinating Committee, the House committee members believed, was being held hostage by CAA's refusal to accede to Army Air Forces and navy demands for GCA. It could do this because the Executive Order establishing the Air Coordinating Committee required unanimity in decision-making. The members demanded that CAA stop using its veto power to block committee decisions and recommended that Congress establish a new body whose structure would not allow CAA to dominate the military. Further, it recommended that the House not fund any new civil landing aids, including the twenty GCA units the army and CAA had negotiated a loan of, until such time as the Army Air Forces, navy, and CAA reached agreement on a common air navigation system. The House agreed, as did the Senate, and CAA found its entire air navigation plan suddenly in ruins.[50] Considerations of national defense thus drove Congress to censure the Civil Aeronautics Administration, which the members were convinced was interfering with the Pentagon's plans.

Yet despite CAA's loss of FY1948 funding, and the harsh criticism it received from the House, Congress's decision actually represented a substantial victory to the beleaguered agency. It had not been stopped from installing and operating the hundred or so ILS systems that had been funded in earlier budgets, nor had it been directed to install GCA in any form. By late 1947, therefore, the major airlines had begun using ILS at thirty-nine airports throughout the nation and continued to do so on more than 1,500 installations in 1998. Despite being castigated in every media outlet from *Aviation News* to *Life* magazine, and being subject of a congressional censure, CAA got what it wanted: an operational ILS system to serve its preferred customers, the commercial airlines.

Congress's nondecision also represented a victory for the commercial airlines, which got a "good enough" system for immediate operation. They were not forced to wait another four or five years for radar landing, which would have happened had Congress forced CAA to conform to the navy's decision. Not incidentally, the airline pilots won, preserving their skills and autonomy for at least a few more years. Their support was probably crucial, since pilots' refusal to use the Hegenberger system in 1934 had doomed that system to failure. Congress certainly listened when Behncke insisted that landing lights were the most important landing aids, for the House final report elevated lights to its own number one priority. Finally, without ALPA's support, CAA would have stood alone. The Air Transport Association, although supportive of ILS, ultimately wanted both systems installed universally, an outcome CAA did not support.

The two losers in this controversy were the U.S. Navy and private aviation. The navy did not lose much. Although its GCA-only policy did not get a congressional mandate, that hardly mattered. Just as the navy's unique requirements had kept it out of the discussion in the RTCA's landing aids effort during the late 1930s, its needs clearly made it nearly irrelevant here. No one expected naval considerations to have great impact on civil aviation. Congress was more interested in the AAF's opinion, which, as the *AOPA Pilot* cartoon showed, was mixed, and therefore open to interpretation. In the light of widespread media support for GCA, the members chose to believe that the army, like the navy, preferred GCA.

The real loser, then, was private aviation. Without money, regardless of its intentions, CAA could not provide any kind of radar landing service to private fliers for at least several more years, and small plane owners had to wait until the transistor made receivers small enough and light enough to use ILS. Administrator Wright, who was personally a champion of private aviation and drove the development of a simplified plane designed for the "average" private flier, resigned a few months after the congressional investigation ended, perhaps frustrated that he could not find a way to please both commercial and private aviation. Private fliers were thus locked out of the air navigation system during poor weather, giving the lie to Wright's own dreams of "commuting in the modern manner" via a plane in every garage.

CONCLUSION

GCA and ILS had politics built in to their very operation. The pilot-control model, on which ILS operated, favored skilled, experienced pilots who flew frequently, making it the obvious choice of professional pilots. The ground control model, on the other hand, favored the occasional pilot, leading to unswerving devotion from advocates of "democratic" flying. The two models sprung not from political differences between the inventors but from their design environment. Diamond and Dunmore, originators of the pilot-control model, worked closely with professional pilots in a project driven largely by the demands of airlines for regularity of service. That their system continued to be the choice of commercial aviation groups is unsurprising. The ground-control model evolved from Luis Alvarez's experiences as a private pilot and a cyclotroneer, while he was working to overcome a military problem. That a military system was the more "democratic" may seem surprising, but the drafted and hastily trained armed forces of World War II had different technical demands than the post-Vietnam professional military. Alvarez's system had fit the needs of a draftee military nicely, and the very qualities that made it useful in that circumstance made it the ideal solution for private aviation.

Just as the way the two systems worked influenced how they were perceived in the political realm, the way they allocated costs impacted how they were seen in Washington's bureaucratic realm. GCA, while less expensive from the perspectives of the navy and private fliers, threatened very substantial additional costs to the Civil Aeronautics Administration. Its leaders had no reason to believe the agency would be given the additional funds to acquire and staff the new equipment, and the agency's budget history suggested that it would be forced to terminate some of its existing functions to provide operators for GCA sets. That is, after all, what had happened when it tried to obtain a small amount of money to man the three test sets it had borrowed from the army. The agency quite rightly believed that GCA, whatever its benefits to aviation, would not receive the financial support necessary to operate it successfully; as political attention began to shift away from landing aids to traffic control over the next several years, the agency's essential poverty became very clear. For CAA, if not for aviation, ILS was the correct choice.

The House investigation, finally, proved only a brief embarrassment to the administration. Congress's decision to zero-fund landing aids handed Harry Truman plenty of ammunition to shoot back with. Congress's nondecision also inspired widespread derision and disgust in the media. Truman's own investigative committee, chaired by John Landis, chairman of the Civil Aeronautics Board, was thus able to turn the tables on his opponents rather easily. That group, picking up on the Army Air Forces' program to integrate the two systems, answered Congress's censure over the next few months.

Transformations

Congress's smacking the Civil Aeronautics Administration (CAA) with the budget axe after the air safety hearings was not quite the final act in the bitter debate between advocates of instrument landing system (ILS) and enthusiasts for ground control approach (GCA). After a new series of airline accidents in June 1947, the administration turned the tables on Congress. Truman convened a Presidential Special Board of Inquiry on Air Safety, chaired by John Landis, chairman of the Civil Aeronautics Board. It published an "interim" report in July 1947 that claimed 35 percent of all fatal accidents in 1946 occurred during approach and landing. Congress' deletion of landing aids from the 1948 budget, the group wrote, guaranteed that "the tragic pattern of the past will inevitably repeat itself."[1] Unless funding for CAA's landing aids program was restored, the report argued, several more years of unnecessary accidents would occur.

Landis's report also attacked Congress for its continued belief that ILS and GCA were mutually exclusive competitors, despite the testimony of respected Army Air Force experts that they were, in fact, complementary. The report further undermined Congress' position by explaining work being done jointly by the AAF, CAA, and the navy at a former navy airfield in Arcata, California, to improve and integrate ILS, GCA, and two other landing aids into a coordinated system to resolve, finally, the blind landing problem. This was, it pointed out "in full accord with the program recommended by the Radio Technical Commission for Aeronautics."[2] The president's board, in short, implied that the great landing aids debate had entirely missed the point. There was no conflict between AAF and CAA policy. Both sought integration of existing landing aids, not a choice of one over the other, and were doing so through a known, respected, and, above all, politically neutral organization.

Recognition of the need for integration of ILS and GCA derived from serious operational problems that the Army Air Forces, and to a lesser extent CAA, had encountered during World War II. Although both systems were successful as landing aids, that success was circumscribed. They did not solve the blind landing problem. Worse, both proved to have limited capacity. They could be trusted to land aircraft safely as long as the landing rate was kept relatively low, but neither could safely handle high-density traffic.

These two operational problems, then, drove the two agencies to reconsider fundamentally the entire issue of landing aids. They faced two political problems, however, that made it difficult for them to promote an integrated, common system. The partisan conflict between Republican Congress and Democratic administration and the equally fractious conflict between the ILS advocates and the GCA enthusiasts both had to be overcome to get funding restored. To resolve these problems, CAA, AAF, and the navy decided via the Air Coordinating Committee to turn the issue over to a revived Radio Technical Commission on Aeronautics (RTCA), which produced a template for not only a landing aids solution but for a comprehensive, nationwide system of air traffic control. Solving the blind landing problem, ultimately, led to the recognition of a new national-scale problem: airport traffic control.

COMPLEMENTARITY: OPERATIONAL IMPERATIVES
FOR INTEGRATION

During World War II, Army Air Forces leaders had believed that solving the blind landing problem, or even the lesser difficulty of the instrument low approach, would lower their high accident rate and reduce the number of days during which air power was grounded. The Royal Air Force's Bomber Command did not fly missions if their meteorologists predicted the ceiling over England to be less than 1,000 feet during the return flight, and the Eighth Bomber Command did not fly if the return ceiling was expected to be less than 500 feet.[3] This meant that during the war, about half of the missions that had to be scrubbed due to weather were canceled not by weather conditions over the target but over England. That was why physicist David Langmuir had told Lee DuBridge of the MIT Radiation Lab in 1943 that a blind bombing radar could only solve half the weather problem.[4] Some sort of landing system had also been necessary to resolve the other half of the bombing halts.

But the 1930s conception of blind landing was insufficient. Even achievement of truly blind landings would not have resolved Eighth Air Force's weather-related accident problem because of midair collisions during bad-weather operations. Eighth Bomber Command lost hundreds of bombers to accidents during the war, many due to midair collisions.[5]

The Army Air Forces did not get around to trying to resolve this new problem until early 1945, when it began putting pressure on the MIT Radiation Lab to help it develop a traffic control radar. After brief resistance by the acting director of the Radiation Lab, F. Wheeler Loomis, to becoming involved in a new project so close to the Lab's expected termination, he agreed to help the Army Air Forces' Watson Laboratories design a new air traffic control radar based on a heavily modified GCA search radar. This was the CPN-18 project, and George Comstock, who had taken over the GCA project after Luis Alvarez and Lawrence Johnston had fled Cambridge for Los Alamos, was put in charge until the Lab's closure. The AAF later selected General Electric as the contractor.[6]

The idea that radar could be used to control aircraft, of course, was not new. The principle was used throughout the war to direct fighters to their targets in a procedure the United States learned from the British, ground controlled intercept. It was also the basis of Alvarez's GCA system, whose search radar was modified into the new traffic control set. The need for a purpose-designed traffic control radar, in turn, derived from a number of characteristics GCA's search radar had that substantially reduced its effectiveness for traffic control.

Alvarez's search radar suffered from several drawbacks. The first was its beam pattern, which topped out at 4,000 feet.[7] The radar could not detect aircraft at higher altitudes reliably, and a traffic control radar certainly needed to detect aircraft at normal operating altitudes of more than 10,000 feet. The war had also seen the development of aircraft capable of operating at up to 30,000 feet, and those aircraft were expected to spawn commercial variations immediately after the war. A traffic control radar had to be able to detect all the traffic in an airport's area if it were to help prevent collisions, and hence it had to scan much higher altitudes effectively.

GCA's search radar also did not provide its operator with altitude information. Height finding had been left out of the search radar to simplify the equipment and because it was unnecessary to the search radar's function of directing aircraft into the precision radar's beams. The accuracy of the barometric altimeter all aircraft had was good enough to get a plane into the precision beams, as long as it could be directed accurately in azimuth and distance. That meant, however, that aircraft at two different altitudes, but the same bearing and range, appeared to the search radar operator as a single aircraft. Further, without altitude information, the radar could not be used for stacking aircraft in holding patterns, which was how aircraft were managed during peak traffic periods. GCA's search radar could thus be used to separate aircraft by distance but not by altitude, a serious drawback in managing a three-dimensional traffic structure.

The CPN-18 project sought to overcome these major issues, while Glen Gilbert, of CAA's Air Traffic Control Division, reported on the presence of other problems that perhaps were more important to CAA. After the tests Gilbert ran at Bryan, Texas, in February 1945, he arranged another set of tests at Fort Dix, New Jersey, that May. The objectives were to evaluate the use of radar as a traffic control aid in separating aircraft, and as an aid in providing simultaneous navigation and let-down guidance. Gilbert found that the system had several flaws that had to be fixed to make it truly effective as an air traffic control radar, including elimination of ground clutter and providing aircraft identification. Identification was particularly important to CAA, so that operators could quickly identify aircraft in danger of collisions and issue warnings. Gilbert's assessment otherwise mirrored the AAF's, with only the need for aircraft identification on the radar display significantly different.[8]

CAA administrator Theodore Wright realized that at least for high-traffic airfields, an air traffic control radar might be financially justifiable, and he took steps both to get suitable radars from surplus military stocks and to get a specification written and circulated for a radar designed specifically for the job, based upon Gilbert's evaluation of GCA. Specification CAA-743, for a commercial air traffic control radar, was the result. The specification was sent to various electronics companies in March 1946, and Gilfillan Brothers, manufacturer of GCA, responded in May. In his letter, the vice president of the company stated that he could not in good conscience submit a bid on a project so similar to the Army Air Forces' CPN-18. He instead proposed modifying some of the wartime GCA sets to serve until the CPN-18 project reached production in a projected 2.5 years. In his proposed modification the number of radar displays would be reduced to two and placed in a tower instead of in a trailer.[9] This was the same modification eventually made to the three AN/MPN-1 sets the AAF loaned to CAA for use at Chicago, La Guardia, and National Airport.

This was the most positive response CAA received to its specification. Other companies submitted bids ranging from $400,000 to $600,000. By comparison, General Electric's bid on the CPN-18 contract had been $1.1 million. Because its FY1947 research and development budget was much less than these figures, CAA had no money with which to begin a radar program. To ensure that his agency was fiscally positioned to deploy an air traffic control radar once one was developed, Wright requested funds for fifty installations in his FY1948 budget submission to the Bureau of the Budget, but that was cut to twenty-five.[10] Radar in any form was not going to reach CAA except as handouts from the wealthier armed services until at least 1949, and the best Wright could do was to take steps to ensure appropriate liaison between the Army Air Forces and his organization in hopes that CAA could influence CPN-18's development.

The CPN-18 project represented a first attempt to replace the AN/MPN-1 GCA search radar with a purpose-built air traffic control radar, capable of handling the high traffic density of major commercial and military fields even in bad weather. The 1930s version of air traffic control, towers equipped with people wielding binoculars, had been enough to deal with high traffic demands during good weather, but faced with all-weather operations demands by both the AAF and the commercial airlines, the older technique proved thoroughly inadequate. Yet just any old sort of radar would not necessarily function adequately, either. Instead, the AAF and CAA determined that a new radar, based on GCA's search radar, was required.

While the air traffic control radar was under development, the House of Representatives' air safety investigation heard from Army Air Forces witnesses on the closely related subject of landing aids. Other witnesses, and most of the questioning, focused on arguments about which of the two landing aids was better, but as the president's Air Safety Board later pointed out, other testimony argued that question was beside the point. Most of that testimony came from Maj. Gen. Harold McClelland, still the Air Forces Communications Officer, and former brigadier general Milton Arnold, head of the Air Transport Association's engineering section.

General McClelland, the officer who had requested the Valentine's Day GCA test in 1943, told the House committee that integration of all useful equipment into a single system was the proper solution to the blind landing problem. McClelland meant that air navigation and control aids had to be considered as parts of a system in order to resolve the operational problems the AAF had encountered during the war. Each individual aid had to be considered in light of its contribution to the whole. Radar was most important for its traffic control uses, while ILS was useful for its easy adaptability to automatic approach and its higher capacity. Since neither ILS nor GCA was a complete approach and landing system in and of itself, the AAF, he told them, was developing two additional parts of an integrated system: runway lights and a British invention called FIDO. McClelland's testimony was seconded by Milton Arnold, who related his experiences in

Europe, before laying out an expansive plan for installing a comprehensive set of approach and landing aids, including the equipment the AAF had under development in its All Weather Flying Program. His proposal ran to $36.65 million for the 160 busiest airports in the United States.[11]

In late 1944, the Army Air Forces had established the All Weather Flying Program to overcome what was rapidly becoming a greater danger to the AAF than its human enemies, the weather. Originally stationed at Clinton County airfield, near Wilmington, Ohio, the program moved to Lockbourne Army Air Base and then back to Wilmington. The moves hindered it, and it did not achieve a great deal before 1946, when it finally resettled in Wilmington. In early February 1946, Maj. Gen. Curtis Lemay, then head of the Army Air Forces Research and Development Board, called a weeklong conference of both civil and military officials to discuss all available blind flying and landing equipment. At that meeting, the army, navy, and CAA agreed to establish a landing aids experiment station to thoroughly evaluate an *integrated* set of approach and landing aids under real world conditions. Because the AAF's All Weather Flying Program operated out of a field with insufficiently poor weather, the Landing Aids Experiment Station was to be established at the former Naval Auxiliary Air Station at Arcata, California, known as the airfield with the worst weather in the forty-eight states. Initially, the navy administered the station while looking for a suitable contractor to run it. A brief contract with the University of Southern California failed in mid-1946, and the navy re-let the contract to United Airlines the same year.[12]

The Arcata Landing Aids Experiment Station (LAES) was an attempt to overcome the inadequacies of both ILS and GCA. Each was effective for the relatively limited application of dealing with single or very few aircraft. For large numbers, however, neither system was really satisfactory. ILS had no means of coordinating the movement of aircraft around an airfield, and that major deficiency resulted in midair collisions among bomber formations during poor weather. ILS needed a traffic control radar to be effective. Further, with no way to provide positive distance separation between approaching aircraft, planes had to be separated in time. That meant one plane had to land before the next one could be allowed to leave its holding pattern and begin an approach. That reduced the theoretical ILS handling capacity of one aircraft per minute to an actual rate of one every ten to fifteen minutes, depending on the airport.[13]

GCA had its own problems. As already discussed, the search radar's capabilities were not adequate to serve as a good traffic control radar. The precision radars, in turn, provided a glide path that was too short for heavy aircraft, and lengthening it meant substantially reducing its handling capacity because the final controller could handle only one aircraft at a time. Lengthening the approach from two miles to ten miles meant the controller had to work with each approaching aircraft for five times as long. Adding operators could increase the capacity, but it also raised the cost. The ILS had no such problem since once on the beam, pilots could fly themselves in. One goal of the Arcata project was, therefore, to develop procedures for using the two systems as complements to each other.

The other major goal was to evaluate two other, nonelectronic systems intended to address the reality that neither ILS nor GCA was an effective blind landing system. Each was an effective approach system, capable of bringing aircraft down to altitudes of 100 feet or so, but although both systems had achieved complete "zero-zero" landings, they could not achieve that safely with every landing. The problem was runway alignment. Both systems could accurately put planes directly above the runway, but could not guarantee that the plane was aligned to go straight down the runway once the wheels touched. Pilots had to verify aircraft alignment visually, which meant operationally that they had to see something outside the cockpit from an altitude of 100–200 feet, depending on the size of the aircraft, in order to achieve alignment before touchdown or to abort the landing.

The two existing systems were FIDO and lights. FIDO was a British development, first put into the RAF Coastal Command base at Davidstowe Moor, the same place Alvarez's prototype GCA was located under Arthur C. Clarke's care. It was simply two lines of large oil burners, one down each side of a runway. The idea was to burn off fog, creating an artificial ceiling of 200 feet or so above the runway. FIDO burned approximately 100,000 gallons of fuel oil per hour, making it a very expensive proposition, but it did remove fog from the field. The U.S. Navy installed one at Amchitka, Alaska, during 1944 for tests but found the base too remote for effective testing after the war. At Arcata, FIDO's fuel consumption was reduced by half by improving the combustion efficiency of the burners, but that still-high expense was compounded by the violent turbulence that the rising heated air generated. Clarke reported that the turbulence was responsible for at least one accident during the wartime British tests. Although large commercial aircraft would be less affected by turbulence than smaller planes, the combination of cost and turbulence made FIDO a commercial failure. One commercial installation was made at Los Angeles in 1949, but no more followed it.[14]

FIDO would have been an effective emergency landing aid, but commercial aircraft could always go somewhere else less fogbound in the 2.5 percent of the time that ceilings were below 200 feet at a particular field.[15] Maintaining an expensive emergency system thus made little sense to the airlines. There was little danger of an expensive crash without FIDO, given the prevalence of alternative airfields in the United States. Passenger discomfort at landing between two walls of flame was no doubt another reason the airlines eventually rejected FIDO.

More controversial than FIDO was the issue of approach and landing lights. Before World War II, both AAF and CAA believed strongly enough in "blind" landings assisted by radio that they simply did not bother investigating the possibility of using lights to improve landing safety. In the late 1930s, the pilots' union (ALPA) had begun pushing for approach lights of some sort, and CAA had grudgingly begun looking at neon "approach light lanes" to provide some visual guidance, but that program got lost in the war fever. Actual runway lights were not explored at all. During the war, the AAF used an RAF lighting scheme that was nothing more than relatively dim lights around the airfield edges called perimeter lights; these were simply there to inform a pilot that he was over the field. That was the total extent of lighting investigations until wartime experience made it brutally obvious that an appropriate system of lights could make a big difference in helping pilots get that critical bit of visual alignment information before landing.

The president of the pilot's union fired the first public round in the lighting controversy, which eventually managed to surpass the ILS/GCA controversy in bitterness.[16] In a February 1946 editorial, he argued that approach and runway lights should be the first priority of the postwar CAA but was soundly ignored by CAA through 1947. He brought it up again before the House Commerce Committee in his testimony during the air safety hearings and was rewarded with placement of lights at the top of the committee's own priority list, although lack of agreement over what kind of lights kept Congress from providing funds for large-scale procurement. The basic conflict was over CAA's intention to install 1,500-foot strips of neon lights in airport approach zones, while the air force, navy, the Air Transport Association and the pilot's union wanted 3,000-foot strips of high-intensity lights. However, the lights issue did not reach the level of true public controversy until the Arcata project results began to appear in late 1948, perhaps because the aeronautical press could not find a way to make lights as compelling an issue as radar had been.

FAA historian John Wilson deals the lighting controversy in his official history of CAA, and it bears mention here only to report the denouement of the Arcata project. Arcata tests demonstrated the superiority of a runway lighting arrangement called "slope-line lighting" during 1948, after which the air force, navy, CAA, Air Transport Association, and the British and French governments all agreed to it. The pilots union refused to accept it, however, and in a true tour de force, managed to defeat slope-line and get adopted its own center-line lighting system in 1950.[17] The pilots union thus humiliated nearly everyone, and in a fit of pique, Congress ordered the Arcata project shut down. The members saw no point in funding experimentation to determine the best methods if the results were simply going to be ignored. Nonetheless, the union got what it wanted over the next several years: the addition of a standardized lighting arrangement at every runway served by an ILS system.

Ultimately, then, the Arcata project had mixed results. Although it was able to reduce FIDO's voracious fuel appetite and to demonstrate the best lighting configuration, neither of those technical successes translated into widespread use. Nor did the project add to the aviation community's knowledge of air traffic control issues or of the performance of ILS and GCA under real-world conditions. Both systems had already been thoroughly lab tested, and the Landing Aids Experiment Station was certainly a less realistic operating environment than that available at the many ILS and GCA installations already in operation at commercial and military airfields. It did, however, represent the first integration of the now-common parts of our airport traffic control systems: surveillance radar, ILS, and approach and landing lights.

The aviation agencies that had established Arcata station and its integrated approach to solving the blind landing problem thus had operational reasons for it and CPN-18 traffic control radar projects. The need to speed traffic near congested airports and prevent midair collisions in poor weather was sufficient to justify adoption of expensive installations of the integrated system, at least in high-traffic areas. But the operational considerations that had persuaded the experts had not been persuasive enough to overcome Congress' predisposition to consider ILS and GCA as competitors, as the outcome of the air safety hearings demonstrates. The aviation community, therefore, had to find a way to overcome the political fallout from the controversy.

POLITICAL IMPERATIVES FOR INTEGRATION

The press attention given to the ILS/GCA controversy during late 1946 and early 1947 had convinced House investigators that CAA and Army Air Forces were pursuing different landing aids, which would impair national defense by making civil airfields unusable by military aircraft. That the two agencies' plans were not actually incompatible was irrelevant. As long as members of Congress believed that the two agencies disagreed, there existed a political issue that the administration could use against the Republican Congress (as Truman's Presidential Air Safety Board had done) or Congress could use against the administration

(as the House Interstate Commerce Committee had done.) The landing aids controversy thus reached the highest level of politics in the United States. Resolving
the 1947 controversy, in turn, meant that CAA and the Army Air Forces, two executive branch agencies, had to convince Congress that they no longer disagreed
and that their proposed solution was not merely a political ploy by the administration intended to embarrass Republican congressional leaders. The controversy
made integration more difficult to achieve, then, because even if the two agencies
agreed on a plan, it was still subject to partisan politics unless some means could
be found to present it in a nonpartisan format.

Even if CAA and the army agreed on landing aids, moreover, it was very clear
that one group was still strongly opposed to CAA's plans. Despite the pious claims
made by Administrator Wright and the chairman of the Civil Aeronautics Board,
John Landis, during their congressional testimony that "all aviation interests" had
agreed to CAA's plans, the controversy revealed that one aviation interest had not
been consulted: the Aircraft Owners and Pilots Association. Since AOPA represented private aviation, and private pilots made up the vast majority of all pilots
in the United States, it had considerable political suasion among Congressmen
interested in ensuring that whatever air navigation system was purchased served
the greatest number of people. Hence resolving the ILS/GCA controversy also
meant finding some way to appease AOPA and its private pilots, thereby satisfying demands by various members of Congress concerned with pursuing the
"greatest good" with taxpayer dollars, and, of course, keeping the votes of tens of
thousands of private pilots.

Integration was the obvious solution to both of these political imperatives.
Crafting a program that would provide both ILS and GCA services, as well as approach and landing lights, at virtually all airports and many smaller airfields
would appease AOPA and its congressional supporters, keep the Pentagon happy,
and thereby mollify Congress. Since integration did, in fact, have valid operational imperatives, a plan to deploy air traffic control radars (Airport Surveillance
Radars, or ASR, beginning in 1946), ground-controlled approach radars (Precision Approach Radars, or PAR), ILS, and landing lights was defensible from the
standpoint of operational need.[18] The plan could not be assailed as purely political. The two major potential roadblocks to integration were the cost of the program, which would surely upset fiscal conservatives in both Congress and the par-
simonious Bureau of the Budget, and the reality that Republicans in Congress
might perceive a plan coming from the Truman administration as an attack on
them.

The cost of the program, as one CAA official put it, would have to stand on its own.[19] Since public pressure had forced CAA to alter its own program, Wright came to believe that Congress had to be the final arbiter of whether to fund a more expansive, and expensive, infrastructure. One result of the controversy was that after 1947 CAA felt more able to propose an expensive program than its traditional poverty would have allowed.

CAA was then essentially freed of budgetary restraints in its planning by a decision of the Air Coordinating Committee to allow the Radio Technical Commission on Aeronautics to devise the program.[20] That decision meant the plan would no longer be the exclusive prerogative of the agency, as the AAF and navy were also represented on the commission, but because this was a nongovernmental committee staffed largely by civilians, it had the freedom to design a system without great concern about cost. This is not to say that cost was no object—the Air Transport Association's representatives could be counted on to help rein in costs lest airlines be forced to help pay them—but that CAA's own budget woes could be subordinated to operational imperatives within the commission's work.

More importantly, however, the Air Coordinating Committee's decision to rely on the nongovernmental RTCA effectively removed the entire issue from partisan politics. Since RTCA was widely perceived to be a group of experts in an age in which expertise was still considered objective and politically neutral, whatever their product turned out to be would be acceptable to both the Congress and the administration. Indeed it was. Fully published in January 1948, the report of RTCA's Special Committee 31 was hailed by everyone, immediately adopted by the Congressional Air Policy Committee as its own, and then blessed by the administration, whose own Air Policy Commission published its recommendations too early to jump on the bandwagon.[21] To get around the minor problem that RTCA had no authority, the administration then established the Air Navigation Development Board with a mandate to turn the committee's recommendations into reality.

RTCA's grand plan did not, however, emerge directly from the Air Coordinating Committee's decision. Although the agency had been moribund during the war, it had begun meeting in April 1944 to consider how best to make use of all the new radio-based technologies fostered during the war. The original constitution of RTCA's executive committee, which initially had representatives from AAF, navy, CAA, State Department, ATA, ALPA, Aeronautical Radio, Inc. (ARINC), and the Federal Communications Commission, was altered in early 1946 to add the one major "out" group, AOPA, at that organization's request. This helped resolve the private flier problem. In December 1945, Delos Rentzel, head of ARINC, submitted an agenda item to the executive committee requesting that

RTCA undertake the development of a comprehensive plan to overcome traffic control and air navigation problems.[22] His request, in turn, was based on a letter from the Air Transport Association. RTCA's examination of the problem thus began before the House investigation and the Air Coordinating decision to turn the decision over to them.

Because much of RTCA's attention during early 1946 was devoted to matters concerning coordination with the new Provisional International Civil Aviation Organization, it did not begin meeting to develop an air navigation plan until August 1946. It produced a brief document that reflected an attempt to grasp the complexities of the new technologies but lacked specificity. It did not detail the amount or cost of equipment needed, and it merely suggested that more research and development needed to be done. Nonetheless, this early report included all of the landing aids that were then being tested at Arcata in its recommendations for "approach zone facilities." During 1947, RTCA refined its recommendations carefully, to ensure that the program effectively served private, military, and commercial fliers. Largely mirroring the Air Transport Association's program in its landing aids recommendation, RTCA ultimately recommended an interim target for 1953 of 320 ILS, 150 airport surveillance radar, and 82 precision approach radar installations. It also stipulated that some of the ASR and PAR sets be installed at noncommercial fields so that service was provided to private aviation. In this way, RTCA intended to heal the rift between CAA and the private fliers' organization. The widespread deployment of precision approach radar that the group proposed also ensured that the radar enthusiasm that the ILS/GCA debate had revealed among the public was addressed, while ILS, already widely deployed, remained to satisfy the pilots union and the airlines' desire for a system now, not in some Buck Rogers future.[23] Finally, this resolved the second political imperative that drove integration. By giving each group the particular technology it favored, RTCA effectively bought their agreement. If no dissenting voices echoed in Congress, the controversy would simply vanish.

RTCA's proposal, however, was far more than simply a duplication of the Arcata project at every commercial airport and many private flying fields in the nation. In the strong political winds, the commission had taken a long reach and proposed nothing less than the creation of an integrated, nationwide, air navigation and traffic control system. By interlinking (networking, in modern terms) automatic long- and short-range radar, the group envisioned positive radar control over every mile of U.S. airspace. Further, it foresaw the use of automation to eliminate labor-intensive flight progress strips and flight clearance management, allowing automatic tracking of aircraft from takeoff to touchdown, and eventually from gate to gate. It was an ambitious plan expected to take twenty years to complete. It was also a prescription for a very expensive command research and development program, and on the Scylla of cost and the Charybdis of technocracy the plan eventually ran aground.

CONCLUSION

Solving the "blind landing problem" required overcoming unforeseen difficulties that derived from both operational and political imperatives. No one in the 1930s had envisioned that massive increases in traffic would necessitate not only bad-weather landing aids but also robust traffic control aids. The AAF's experiences during World War II demonstrated that need and further revealed the utter inadequacy of its earlier attempts to beat the weather. By the end of the war, both AAF and CAA realized that more research and development was needed to truly defeat weather constraints on aviation. The operational inadequacies of SCS-51 and AN/MPN-1 meant both systems needed to be supplemented and improved, an imperative the aeronautical press entirely missed. Their different models of operation, pilot control and ground control, proved highly complementary, and their integration into an alliance, albeit an uneasy one, was one foundation of the modern air traffic control system.

The creation of an integrated system of approach and landing aids was not solely driven by operational demands, however. The public frenzy over air safety and radar took the issue out of bureaucrats' offices and put it before a Congress only too willing to grab it, attack the administration with it, and demand a solution. The dizzying political heights to which the ILS/GCA issue ascended in the United States ensured that the radar versus instrument landing question was at least addressed, even if the answer, SC-31, eventually failed of its promise.

Despite congressional and administration approval of SC-31 and allocation of funds for it beginning in FY1949, the next four years saw the program stall. The Korean War caused the diversion of increasing portions of CAA funding to the Pentagon, weakening its ability to influence Air Navigation Development Board research and essentially eliminating CAA's airways equipment program, while the election of Dwight Eisenhower installed a president who, as Walter McDougall has argued, resisted the siren call of technocratic management and command research and development. Without strong leadership from the chief executive, the Air Navigation Development Board's cooperative effort quickly disintegrated into interagency squabbling, and it achieved little. As a result, CAA had a total of ten precision approach radar systems in 1950, and the number never grew larger.[24] The ten airport surveillance radar sets it had installed by 1950 also remained unchanged through 1953, and the number of long-range surveillance radars remained fixed at two—both World War II sets on permanent "loan" from the air force—until another air safety crisis caused Congress to dismantle the ineffectual CAA and replace it in 1958. The resulting new agency, the Federal Aviation Agency, was responsible for finally pulling together the postwar technologies of aviation into the modern air traffic system.[25]

Conclusion

On August 6, 1997, Korean Air flight 801 crashed near the top of Nimitz Hill, on the U.S. territory of Guam, near the transmitting antenna of the visual omnirange (VOR) that aircraft approaching the island's primary airport outside the city of Agana use to maneuver onto the ILS approach path. Immediately, the press turned its attention to the "ancient" ILS installation, which had both of its glide path transmitters out of service. The plane's crew was aware of the outage and had been briefed on the approach procedure for use when the glide path transmission is unavailable, but they did not properly follow it. Ultimately, the National Transportation Safety Board ruled the accident a product of "pilot error." There were, of course, contributing factors.[1] But as one diehard supporter of Luis Alvarez's ground control approach system wrote to the industry magazine *Aviation Week and Space Technology,* this accident would not have happened if GCA (or its civil equivalent, precision approach radar) had been the landing aid chosen forty years before.[2]

So much is certainly true. The final controller of a GCA installation would have been able to warn the plane's crew that they were descending toward Nimitz Hill and not the runway three miles further along their flight path. But a GCA world would have had accidents of its own. Perhaps the most famous GCA failure came amid its most important success, the Berlin airlift of 1948 and 1949. On August 13, 1948, forever after known as Black Friday, the airfield in the U.S. sector of Berlin, Templehof, suffered a series of accidents. One C-54 missed the runway and crashed, while the aircraft behind it blew its tires trying to stop before hitting the wreck. A third C-54 crashed on the auxiliary runway, catching fire and burning up plane and cargo.[3] The relatively inexperienced final controller in Tempel-hof's GCA unit, who would normally have been able to prevent these sorts of accident by proper instructions to the approaching pilots, lost his "nerve," as the euphemism of the day went. As a result of Black Friday, the airlift commander established much stricter traffic control procedures and put in place one of the central rules of modern commercial aviation: all aircraft had to operate under instrument flight rules, regardless of weather conditions. That effectively placed the aircraft under radar control, while ensuring that all pilots followed the same exacting flight procedures.

These two bad days forty-nine years apart indicate that neither ILS nor GCA, ultimately, was perfect. If GCA could have prevented the KAL accident in Guam, some other catastrophe related to GCA's imperfections would have taken its place in the long years since its 1943 debut. What makes the ILS/GCA decision still contentious forty years later is not competing claims for safety, although the dispute is often expressed in those terms, but the question first raised by Rep. M. Harris Ellsworth during the 1947 hearings: how to do the most good for the most people? In other words, whom should the airways be built to serve? That question places the landing aids issue into the "democratic technology" arena in the historiography of technology. What, exactly, a democratic technology might look like is as problematic as the issue of what a democracy is, but that has not stopped scholars from trying to define them.

Among the most prominent advocate of rules to define and regulate technologies is Langdon Winner. Winner proposes a general maxim, that *"technologies be given a scale and structure of the sort that would be immediately intelligible to non-experts."*[4] Acceptable technologies would be limited to those whose operation is obvious to the proverbial "casual observer." That would, unfortunately, exclude both ILS and GCA, as even some commercial pilots cannot explain how ILS works. They only know how to use it. Indeed, Winner's rule would strike out all aviation technologies. Most people cannot explain how aircraft fly and perhaps it bears mentioning here that no one since the Wrights has been able to get a reproduction of the famed 1903 Flyer aloft. If its flight had not been recorded on film, historians would probably have had to reject their claim to the first powered flight because their experiment has proven unrepeatable. If even the Wright Flyer's operation is nonobvious, then there are no simple technologies of flight. But are there no democratic ones either?

Perhaps a more useful way to approach the issue of democratic technologies can be drawn from feminist historiography of technology. Although much of that literature focuses on various aspects of gender and technology, one underexplored aspect deals with issues of access. "Access" means many things to many people, from representation in the design process to physical possession of a technology, so perhaps substitution of a more precise term is in order. By "accessibility" I mean what we might call usability. Alvarez's GCA was clearly more useable to an average pilot than was ILS, as most pilots, congressmen, and aviation writers recognized. One implication of GCA's usability by a broader range of fliers would have been more opportunity to fly for owners of small aircraft. Though political philosophers may balk at the crudity of this analysis, it seems reasonable to assert that a technology whose characteristics make it useable by a larger number of people than its competitors is the more accessible, and the more democratic.

Although Joseph Corn locates the death of the dream of democratic aviation in the late 1940s, the persistence of GCA advocates suggests that the dream is alive and well in AOPA, still the most contrarian organization in American aviation. J. B. "Doc" Hartranft's organization has continued to be a persistent critic of CAA successor, the Federal Aviation Administration. In particular, the organization has vehemently resisted the adoption and maintenance of radar control of traffic around the busiest U.S. cities. Although fans of radar landing, AOPA's members chafe at the complex traffic control procedures that FAA has built around its surveillance radars to manage traffic, and blame those procedures for diminished interest in personal flying among the public. Radar has been, at best, a mixed blessing for its original champion.

As one might expect, FAA was driven to expand the scope and complexity of its radar-based traffic control procedures by the rapid growth of commercial aviation. Because the nation's busiest airports face much greater traffic than did Berlin during the airlift—one landing every three minutes in Berlin meant one landing every nine minutes at each of Berlin's three airports, a trivial rate by modern Chicago-O'Hare standards—the traffic control procedures are relatively more intricate and rigorous. The effect of those procedures, as AOPA has complained for two decades, has been to drive general aviation out of the major airports, reducing its members' access to the skyways. In some sense, that result is a denial of the dream of many of the "air-minded" pioneers of aviation that Corn documented so well, but increased commercial traffic has also meant low enough airfares that nearly everyone can fly, even if the piloting is done by someone else. If the democracy granted by low air fares is not quite what the true believers in democratic aviation intended, perhaps they can be forgiven for not quite grasping the realities of twenty-first-century America: in an increasingly urban society, not everyone can own a plane. The parking issues alone make the idea almost laughable.

Along with more complex rules under which flying is done, aviation has seen a proliferation of interest groups representing specific subsets of the industry. While CAA had to deal only with the Air Transport Association, ALPA, and AOPA in addition to the army and navy during the development of ILS and the ILS/GCA controversy, FAA must now deal with a plethora of what it calls the "alphabet groups." These include, for example, the Experimental Aircraft Association (EAA), representing owners of home-built aircraft; the Allied Pilots Association (APA); the National Air Traffic Controller's Organization (NATCO); and the National Business Aircraft Association.

The proliferation of these groups has created a situation similar to that described by Brian Balogh in his study of the decline of the nuclear power industry in the United States.[5] He refers to the construction of "issue networks" surrounding the industry composed of activist groups focused on various aspects of nuclear plant safety, but in aviation the situation is a bit different. While the groups Balogh studied were oriented around a single issue, and were often ephemeral, the alphabet groups are permanent organizations active in a wide array of issues within aviation. Sociologists, of course, also perceive such networks, redefined as "actor networks," in their studies of the activities of scientists and engineers and the managerial professionals that support them. Actor networks, like the often-ephemeral issue networks Balogh finds, seem too impermanent to explain the longevity of technologies. We should describe the network formed by the alphabet groups, FAA, and the Pentagon's four different air arms as an institutional network. Although sociologists understandably balk at allowing faceless institutions agency, the character of the major aviation organizations has changed very little over time, and their positions on various aviation issues have also been changeless. Individuals are shaped, in part, by the institutions they represent, and the conception of an institutional network allows for personalities to act in that context.[6]

The negotiations between CAA, the airlines, and the Army Air Corps during the 1930s stretched over years due to a combination of differing technical and training needs and misunderstandings. Commercial adoption of a standardized landing aid was retarded by the need to negotiate a standard acceptable to all parties in order to gain their support. FAA is often criticized for its inability to adopt new technologies quickly, and one reason is simply its need to satisfy the conflicting demands of the growing number of alphabet groups. While CAA's ILS ne-

gotiations in 1938 and 1939 involved only itself, the air corps, airline representatives, and the representatives of a few radio companies, FAA now must deal with
a much wider range of interest groups. It no longer has the assistance of RTCA's
extragovernmental committee structure to aid the negotiation process, however.
The Kennedy administration removed the government representatives from
RTCA, believing that allowing industry and government representatives on the
same committee violated conflict-of-interest laws. RTCA still exists as a private
foundation dedicated to aiding the adoption of new technologies into aviation by
fostering communication among the various groups, but it no longer holds the
informal authority it once did.

The Federal Aviation Administration thus faces a more difficult task in developing acceptable standards than did its predecessor. Although FAA's legal authority is much greater than that of its predecessors, it has limits. It cannot simply ignore the alphabet groups because those groups have proven perfectly willing
to wage media campaigns and lobby congressional supporters over disagreements. And the alphabet groups are often successful in their challenges. The development of a complex institutional network has ensured that the alphabet
groups scrutinize every technological decision FAA makes, and aspects of the
technology found wanting by a particular group are subjected to lengthy negotiation.

The composition of the institutional network I have described changed over
time and not simply due to the proliferation of alphabet groups. The relative
power of various groups to affect the industry and its technologies has also
changed. During the 1930s, the demands made by commercial airlines on CAA
and its predecessor, the Bureau of Air Commerce, to deploy networks of weather
stations, radio navigation ranges, and communications stations made commercial considerations the dominant influence in infrastructure development, while
World War II put the Army Air Forces in the pilot's seat. The cold war and the
related national security state kept the air force strong enough to prevent the
reemergence of a dominant CAA during the 1950s, until Congress, via the Federal Aviation Act of 1958, rather surprisingly made the new FAA entirely responsible for navigational infrastructure and air traffic control. That, in turn, ensured continuation of the commercial bias in aviation policy, while the research
and development resources available to the air force enabled that organization to
dominate in new technologies of flight. Like CAA at the end of World War II, FAA
is generally left to modify air force technologies for civil use, or more commonly,
wait for a private company to commercialize the technology, then approve it.

The diversification of the institutional structure of U.S. aviation that began in the 1930s also permanently linked the industry to academic organizations. While the Guggenheim Foundation had funded aeronautical research at several universities during the 1920s, that mode of support for research did not give the researchers a policy voice. Further, aeronautics research, though very important, was also circumscribed. The laboratories founded by the Guggenheim performed extensive research into aerodynamics, but not in electronics, propulsion, radio, or physiology.[7] Edward Bowles had to scrape to get funding for his Round Hill group before CAA took interest in 1937, although MIT possessed one of the Guggenheim labs. The Army Air Corps' wartime needs cemented the university research–aviation linkage, and expanded the scope of government-supported aviation research in universities to include all aspects of flight. Although the MIT Radiation Lab closed after World War II, the air force had grown so used to having its own microwave scientists that it got MIT to establish a new lab to assist it, the Lincoln Laboratory. That organization, in turn, was crucial to FAA's microwave landing system (MLS) development program during the 1970s. MLS was supposed to replace ILS during the 1980s and 1990s, but its financial costs proved more than either administration or Congress were willing to bear. Similarly, the Pentagon's JASON group maintained a policy voice for scientists in civil aviation, perhaps most prominently in advising President Kennedy to build a separate air traffic control system during the 1960s, instead of using the air force's SAGE air defense system.[8]

There are, therefore, tight links between military and commercial aviation, at different levels. They increasingly share the same technologies, while the same laboratories and scientists serve to advise military and civil aviation policy officials. Yet this was not always the case, as Chapters 1–3 suggest. During the 1920s and early 1930s, military and commercial technologies diverged, with commercial airlines demanding runways and greater precision in navigation and landing aids, while the Army Air Corps was satisfied with grass fields and the first generation of radio aids—low-frequency communications, ranges, and compass locators. The cause of the divergence was commercial demands for regularity of service, an issue that did not matter to the interwar Air Corps. Why the reconvergence, then? World War I was fought without runways and radio navigation aids, as Eddie Rickenbacker's description of "dud days" attests, and no one seems to have minded the sometimes days-long absence of aircraft over the battlefields. What made World War II airmen suddenly adopt commercial technologies designed to allow all-weather flying?

Veterans of the Eighth Air Force generally consider the film *Twelve O'Clock High* to be the truest depiction of their war against the Luftwaffe, and perhaps from there we can begin to draw an answer: *maximum effort.*[9] Hap Arnold placed enormous pressure on Ira Eaker to show results for the vast expenditures made by the United States to equip Eighth Air Force with its aerial armada. Without a doubt, that reflected pressure Roosevelt and the joint chiefs placed on Arnold in order to justify expending those resources on bombers and not on ships, landing craft, and tanks. In part, this pressure was a result of Arnold and Eaker's own argument, that strategic bombing could achieve decisive victories early in the war. As Eighth failed to produce those victories, the AAF's field commanders, with their RAF counterparts, tried instead to exert constant pressure on the enemy. Eaker thus sought to fly in conditions that would have seen Rickenbacker's Flying Circus safely grounded in order to produce that maximum effort, and his replacement, James Doolittle, pushed the weather limits still harder. To that end, the Army Air Forces, leaning heavily on commercial technologies, sought to become more like the airlines. All-weather flying (and bombing and air combat) would allow near-continual pressure on the enemy. Radar, once integrated with the commercially inspired radio navigation equipment designed during the 1930s, enabled completion of the "all-weather flying system" the airlines and CAA had begun through blind landing research. The AAF thus adopted a military version of regularity of service during the war as its goal. The motivation of the two groups was different: whereas the airlines sought to become profitable alternatives to the railroads, the AAF sought to defeat the Luftwaffe, justify its budget, and achieve its bid for independence from the army's hated ground forces. For different reasons, then, the AAF and the airlines came to require all-weather flying.

Recent scholarship in the history of science has shown that the Defense Department launched substantial research efforts into atmospheric sciences after World War II, based on use of the German V-2 missile. Understanding of the weapon's operating environment, researchers believed, was necessary to improving its performance.[10] Yet that was, in essence, no more than an extension of the AAF's belated recognition of the importance of environmental considerations in aviation. Weather-related crashes were not the only evidence of the AAF's prewar lack of respect for nature's inconsistency, after all. The unheated, unpressurized B-17 proved murderous to early crews, who could be permanently disabled by frostbite if the Luftwaffe failed to kill them. Gun crew members also sometimes froze to death, while others suffocated on vomit that froze in their oxygen masks.[11] The commercial airlines had recognized the weather as their primary enemy in the late 1920s; the AAF, by not adopting the dream of all-weather flight as quickly as the airlines had, found itself nearly as vulnerable to nature as it initially was to the Luftwaffe. The Pentagon's deployment of vast resources after World War II to understand what its officers perceived as the "operating environment" makes perfect sense, given the experiences its leaders had in fighting the European weather.

By the beginning of the Berlin airlift, then, a remarkable convergence in commercial and military aviation had occurred, centered on the goal of regularity. Knowledge of the operating environment was merely a means to that end. Sharing the same goal did not necessarily mean that these two subsets of aviation shared precisely the same technologies. Although Curtis LeMay adopted ILS as the air force's primary landing aid in 1950 at its permanent installations, the air force has kept more modern versions of the ground controlled approach system in service for deployment to forward airfields because it is much less sensitive to site conditions. In a practical sense, it is simply more portable than the site-sensitive ILS. Hence the differences between military and commercial aviation's operating circumstances continue to influence their technological choices.

Finally, there is the evolution of the concept of blind landing. The concept had considerable suasion and staying power. The technological frame constructed around the technologies of radio blind landing caused its believers to overlook an "old technology" partial solution, lighting. It should not have taken the experiences of World War II to convince aviation officials that some judiciously chosen application of Thomas Edison's invention could help fliers land. But belief in the possibility of blind landings guided solely by radio was compelling enough that even after the war, Charles Stanton and others insisted on retaining the name instrument landing system, still convinced that it could bring about routine blind landings.

Fortunately for the flying public, retention of the name never led to implementation of blind landings using ILS, although experimentation along those lines continued for many years. Much of that experimentation moved into microwave-based landing aids like Sperry Gyroscope's in the quest for greater stability. One system developed during the 1960s for the Lockheed C-141 military transport, the all-weather landing system (AWLS), was supposed to achieve widespread commercial deployment, but it did not. The only true blind landing systems in use today are the microwave landing system designed in the late 1960s that the NASA space shuttle uses, and the navy's carrier controlled approach system, a fully automated version of GCA. MLS was approved for worldwide deployment in 1977 but was canceled in 1995 due, in part, to commercial development of the Pentagon's global positioning system, which promises to cost FAA an order of magnitude less than MLS would have. And although the carrier controlled approach system is installed on all of the U.S. Navy's dozen aircraft carriers, it is never used. The navy's carrier pilots, who are willing to be fired off the deck by big steam-powered pistons, be shot at by enemies of whatever description, and let their returning aircraft get caught by thick cables and sometimes nets stretched across the deck, cannot countenance a blind landing.

In this study, we have seen the devolution of "blind landing" into "instrument approach," and the reason is simply this: although engineers and scientists have produced equipment capable of blind landings, they have not been able to make pilots capable of it. The human element of the system has proven to be the most difficult to engineer properly. The problem is not one of training or skills but one of psychology. Although military and commercial pilots make their living in one of the most technically oriented professions humanity has ever created, they appear to be almost genetically incapable of turning the most dangerous part of a flight over to a machine that, in blind conditions, they cannot monitor. This is the result of that most basic animal instinct, survival. A minor glitch—all too common even in the most modern aircraft—in any other part of a flight is generally an annoyance. A minor glitch during a blind landing is almost certain to be fatal to pilots, passengers, and quite likely people on the ground, too. Hence RAF veteran Frank Griffiths recounts that after several weeks of making blind landings "under the hood" in 1944 using Major Moseley's approach coupler, he and his crew went out to try it for real by making a night approach with field lights and the plane's landing lights turned off. He found the pressure too much, though, and faced with dying if the black box was wrong, he cheated and turned on the plane's landing lights.[12]

The devolution of the concept of blind landing occurred as the technology improved, but more importantly, as knowledge of pilots' reactions to the technologies of blind landing was developed. That knowledge was not formal. "Human factors" research in aviation began under AAF funding only during World War II and remains a hotly contested form of knowledge. Yet there was clearly pilot feedback in the process of design, testing, and negotiation, with the pilots' standard for the "almost-straight" glide path of 1939 the clearest example. The airlines were the first to adopt the stand that blind landings were impractical, likely motivated in part by the repeated refusal of pilots to use the various descendants of the NBS system as landing systems, as opposed to approach aids. The goal of all-weather flying drove the evolution of blind landing systems, but was not enough to overcome the mental hazard pilots were subjected to in a blind landing. Blind landing systems were redefined as instrument approach aids and air traffic control systems in the face of human psychological limitations.

Despite its ultimate failure, the dream of blind landing was thus important to aviation because it stimulated development of an array of technologies that proved important to related needs. Those technologies, in turn, allowed commercial aviation to achieve greater safety and greater regularity than any other transportation system while permitting round-the-clock, almost-all-weather aerial combat and bombardment. But the dream asked too much of the human part of the system. Although Harry Guggenheim announced the achievement of blind landing in 1929, the twentieth century did not see its routinization.

Notes

ABBREVIATIONS

AFHRA	Air Force Historical Research Agency, Maxwell AFB, AL
NA	National Archives, Washington, DC
NAS Archives	National Academy of Sciences Archives, Washington, DC
RG	record group
USGPO	U.S. Government Printing Office

INTRODUCTION

1. Antoine de Saint-Exupéry, *Night Flight* (New York: Century, 1932).

2. Charles Lindbergh, *The Spirit of St. Louis* (New York: Charles Scribner's Sons, 1953), 6–7.

3. "A Classification of the Causes of Forced Landings, from January 1924 to July 1925, with Recommendations," 7 November 1925, file A14.48: Mgr. Monmouth Ill. Development Division, box 47, Records of the Post Office Department, General Correspondence of the Air Mail Service, record group (RG) 28, National Archives, Washington, DC (NA).

4. I'm grateful to my colleague Michael Buckley for the phrase "military-postal-industrial."

5. William M. Leary, "The Search for an Instrument Landing System, 1918–1948," in *Innovation and the Development of Flight*, ed. Roger Launius (College Station: Texas A&M University Press, 1999), 80–99; James R. Hansen, "Aviation History in the Wider View," *Technology and Culture* (1989), 643–56.

6. Nick A. Komons, *Bonfires to Beacons: Federal Civil Aviation Policy under the Air Commerce Act, 1926–1938* (1978; reprint, Washington, DC: Smithsonian Institution Press, 1989), 151.

7. Ibid., 125–46. The best history of the airmail service is William M. Leary, *Aerial Pioneers: The U.S. Airmail Service, 1918–1927* (Smithsonian Institution Press, 1985). Only nine of the Post Office's original forty mail pilots were still alive in 1925; Roger E. Bilstein, *Flight in America: From the Wrights to the Astronauts*, revised ed. (Baltimore: Johns Hopkins University Press, 1994), 52. For comparison of railroad and airline times during the early 1930s, see Edward W. Constant II, *Origins of the Turbojet Revolution* (Baltimore: Johns Hopkins University Press, 1980), 166–67.

8. Robert Wohl, *A Passion for Wings: Aviation and the Western Imagination, 1908–1918* .
(New Haven: Yale University Pess, 1994), quote from p. 1.

9. Joseph Corn, *The Winged Gospel: America's Romance with Aviation, 1900–1950* (New
York: Oxford University Press, 1983), vii.

10. Ibid., 46–50.

11. See Tom D. Crouch, "An Airplane for Everyman: The Department of Commerce and
the Light Airplane Industry, 1933–1937," in Launius, *Innovation and the Development of
Flight*, 166–87.

12. Corn, *Winged Gospel*, 139–40; Komons, *Bonfires to Beacons*, 68. Boeing Air Transport merged with National Air Transport, Varney Air Lines, and Pacific Air Transport to
form the core of United Air Lines. See Frank J. Taylor, *High Horizons* (New York: McGraw-
Hill, 1964).

13. The best synthesis of strategic bombing fantasies is Michael S. Sherry, *The Rise of
American Airpower: The Creation of Armageddon* (New Haven: Yale University Press, 1987),
1–9. The classic histories of strategic bombing are Raymond H. Fredette, *The Sky on Fire:
The First Battle of Britain, 1917–1918* (New York: Harcourt, Brace, Jovanovich, 1976) and Lee
Kennett, *A History of Strategic Bombing* (New York: Scribner, 1982). On Mitchell's struggle
to militarize aviation, see Sherry, *Rise of American Airpower*, 29–31, 34–37, and Bilstein,
Flight in America, 43. A full length treatment is Alfred F. Hurley, *Billy Mitchell: Crusader for
Air Power* (Bloomington: Indiana University Press, 1975). John R. M. Wilson, *Turbulence
Aloft: The Civil Aeronautics Administration Amid Wars and Rumors of Wars, 1938–1953* (Wash-
ington, DC: Federal Aviation Administration, 1979), 63–83.

14. Eric Schatzberg, *Wings of Wood, Wings of Metal* (Princeton: Princeton University
Press, 1999), 5, 15–19.

15. William Cronon, *Nature's Metropolis: Chicago and the Great West* (New York: W. W.
Norton, 1991), xix.

16. An Army Air Forces C-54 cargo plane rigged with a special autopilot system flew
nonstop from Stephensville, Newfoundland, to Ireland in 1947. See "Automatic Controls
for Pilotless Ocean Flight," *Electronics* (December 1947): 88–92.

17. Erik M. Conway, "Echoes in the Grand Canyon: Public Catastrophes and Technologies of Control in American Aviation," *History and Technology* 20, no. 2 (June 2004): 115–
34.

CHAPTER 1: INSTRUMENTAL FAITH

1. Monte Dwayne Wright, *Most Probable Position: A History of Aerial Navigation to 1941*
(Lawrence: University Press of Kansas, 1972).

2. See ibid., 70, for a detailed description of the magnetic compass's airborne misbe-
havior.

3. Annual Report of the Advisory Committee for Aeronautics, 1915–1916, *Flight* (24 August 1916): 720–22.

4. See Elmer Sperry Jr., ts. dated December 1969, Elmer Sperry Papers, National
Air and Space Museum Archives, Washington, DC; description based on U.S. Patent
1,407,491, Elmer A. Sperry, "Turn Indicator," 21 February 1922. Pioneer Instrument Com-

pany memos on the patent issues involved make clear that the idea for the Sperry Turn Indicator came from Luis de Florez of the Navy Bureau of Construction and Repair (during World War II, de Florez became head of Navy flight instrument research), while Charles H. Colvin, who left Sperry Gyroscope to help found Pioneer Instruments in 1919, actually perfected the turn indicator. Elmer Sperry did not initially believe the turn indicator would work. See M. M. Titterington to Lt. E. W. Rounds (Aviation Section, Bureau of Construction and Repair), 9 August 1921, Elmer Sperry Papers. Historian Thomas Hughes gives a different account of the turn indicator in *Elmer Sperry: Inventor and Engineer* (Baltimore: Johns Hopkins University Press, 1971), 233–36. "Le Gyrorector," *L'Aérophile* (1–15 Janvier 1930): 16.

5. Charles Colvin, interview, 27 May 1971, p. 22, Lawrence B. Sperry Papers, National Air and Space Museum Archives; Elmer Sperry Jr., ts. dated December 1969, Elmer Sperry Papers.

6. Quote from Elmer Sperry to Lawrence Sperry, August 10, 1920, Lawrence Sperry Papers. On Alcock's flight, see Sir John Alcock and Sir Arthur Whitten Brown, *Our Transatlantic Flight* (London: William Kimber, 1969). William Davenport, *Gyro! The Life and Times of Lawrence Sperry* (New York: Charles Scribner's Sons, 1978), 253. Apparently, the engine on his Sperry Messenger failed and he glided in. A coastguardsman saw the airplane go down and a search found the plane two hours after the crash about two miles from shore, but Lawrence was never found.

7. Elmer Sperry to Lawrence Sperry, 10 August 1920, Lawrence Sperry Papers.

8. The best history of the U.S. Post Office's airmail service is William Leary, *Aerial Pioneers: The U.S. Air Mail Service, 1918–1927* (Washington, DC: Smithsonian Institution Press, 1985).

9. Ibid., 87–89.

10. Ibid., 171–85.

11. Ibid., 205.

12. Eighteen pilots died in the first thirty-four months of Post Office airmail service, fully half the original complement of pilots. Ibid., 95–112, 147.

13. Ibid., 86–87, 109, 188–90.

14. Ray H. Boudreaux, "The Ocker-Meyers Method of Blind Flying," *Aero Digest* (July 1928): 48, 183–85; William C. Ocker, "Economic Value of Flying by Instruments," *Aero Digest* (October 1930): 62–63.

15. William C. Ocker and Carl J. Crane, *Blind Fight in Theory and Practice* (San Antonio, TX: Naylor Printing Company, 1932), 7.

16. A German instrument called the "Gyrorector" existed that was a workable artificial horizon, but it appears not to have been marketed in the United States. The army obtained one via a military attaché in Europe and tested it at Wright Field in 1926 but did not find it suitable for military aircraft. Its weight also discouraged airline use.

17. William C. Ocker, "Under the Hood," *Southwestern Aviation* (May 1934), 4–6, 28.

18. Ocker and Crane, *Blind Fight*, 12; I use the definitions of "natural" and "mechanical" pilots described by Ocker supporter Boudreaux in "The Ocker-Meyers Method," in this analysis (see n. 14).

19. Hughes, *Elmer Sperry*, 234.

20. "Rosenbaum Gyrorector," War Department report, 28 May 1926, ts., Turn and Bank Instruments Technical File, National Air and Space Museum Archives.

21. Hughes, *Elmer Sperry*, 235.

22. Richard Hallion, *Legacy of Flight: The Guggenheim Contribution to American Aviation* (Seattle: University of Washington Press, 1977), 101–27; *Solving the Problem of Fog Flying* (New York: Daniel Guggenheim Fund for the Promotion of Aeronautics, 1929); Hallion, *Legacy*, 117.

23. Elmer A. Sperry Jr., "The Artificial Horizon," *Aeronautical Engineering* (October–December 1930): 289–91.

24. Hallion, *Legacy*, chap. 6; James H. Doolittle, "Flying an Airplane in Fog," *SAE Journal* (March 1930): 318–20, 345.

25. Wolfgang Langewiesche, "Flying Blind," *Harper's Magazine* (April 1947): 328.

26. "Blind Flying and the Airlines," *Aviation* (August 1932): 349. See also Harold C. Stark, *Instrument Flying*, rev. ed. (Pawling, NY: Harold C. Stark, 1934), iv.

27. Stark, *Instrument Flying*, 6, 8.

28. Ibid., 15. The same system was published by two army fliers the following year under the name "A-B-C system." The army fliers' book is William C. Ocker and Carl J. Crane, *Blind Flight in Theory and Practice* (San Antonio, TX: Naylor Printing, 1932).

29. The Air Corps introduced instruction in June 1930. See "Blind Flying and the Airlines," 350.

30. E. A. Cutrell, "Instrument and Radio Flying," *Aviation* (June 1935): 11–14; E. B. Schaefer, "The Stark System of Instrument Flying," *Aviation Engineering* (March 1932): 21–22; Karl S. Day, *Instrument and Radio Flying* (Garden City, NY: Air Associates, Inc., 1938), vi.

31. "Blind Flying and the Airlines," 349–52.

32. Day, *Instrument and Radio Flying*, 1.

33. Private pilots had no minimum number of solo hours to receive a private license at this time. Komons, *Bonfires to Beacons*, 97; Luis W. Alvarez, *Adventures of a Physicist* (New York: Basic Books, 1987), 30.

34. Robert B. Parke and Lloyd L. Kelly, *The Pilot Maker* (New York: Grosset & Dunlap, 1970): 3.

35. Ibid., 18–20, 26.

36. Ibid., 36.

37. Ibid., 38.

38. Leary, *Aerial Pioneers*, 224–25.

39. The best account of Brown's activities is F. Robert Van der Linden, "Progressives and the Post Office" (Ph.D. diss., George Washington University, 1996). See also Komons, *Bonfires to Beacons*; Leary, *Aerial Pioneers*, 222–37; Frank J. Taylor, *High Horizons* (New York: McGraw-Hill, 1964); and Robert J. Serling, *Eagle: The Story of American Airlines* (New York: St. Martin's, 1985), 63–76. F. Robert van der Linden, *The Boeing 247: The First Modern Airliner* (Seattle: University of Washington Press, 1991), 20–21; Taylor, *High Horizons*, 73–94; Komons, *Bonfires to Beacons*, 197–210.

40. This account is a very brief synthesis of sources that do not all agree on how, exactly, Roosevelt came to cancel the contracts, although they do agree on why. See Komons,

Bonfires to Beacons, 219–99; Norman E. Borden, *Air Mail Emergency 1934* (Freeport, ME: Bond Wheelwright, 1985); and Benjamin Foulois with Col. C. V. Glines, *From the Wright Brothers to the Astronauts* (New York: McGraw-Hill, 1960): 235–59.

41. Komons, *Bonfires to Beacons*, 261–62, 272–73.

42. Parke and Kelly, *Pilot Maker*, 52–53.

CHAPTER 2: PLACES TO LAND BLIND

1. Deborah G. Douglas, "The Invention of Airports" (Ph.D. diss., University of Pennsylvania, 1996), 322–23.

2. John Law has argued for the importance of understanding "natural" factors in the design and success of technological systems. See his "Technology and Heterogeneous Engineering: The Case of Portuguese Expansion," in *The Social Construction of Technological Systems,* ed. Wiebe E Bijker, Thomas P. Hughes, and Trevor Pinch (Cambridge, MA: MIT Press, 1987), 111–34.

3. Maj. Gen. Benjamin D. Foulois, with Col. C. V. Glines, *From the Wright Brothers to the Astronauts* (New York: McGraw-Hill, 1968), chap. 7.

4. Douglas, "Invention of Airports," 275–77.

5. Capt. Eddie V Rickenbacker, *Fighting the Flying Circus* (Garden City, NY: Doubleday, 1965), 9–10, describes World War I French aerodromes.

6. A. C. Blackall, "Why Britain Uses Grass for Runways," *Airports* (September 1928): 38; Hans-Joachim Braun, "The Airport as Symbol: Air Transport and Politics at Berlin-Tempelhof, 1923–1948," in *From Airships to Airbus: The History of Civil and Commercial Aviation, Vol. 1: Infrastructure and Environment,* ed. William Leary (Washington, DC: Smithsonian Institution Press, 1995), 45–54.

7. H. Oakley Sharp, G. Reed Shaw, and John A. Dunlop, *Airport Engineering* (New York: John Wiley & Sons, 1944).

8. "Addendum to Air Service Information Circular, vol. IV, no. 303—Discussion of Airplane Tires and Wheels" (Washington, DC: USGPO, 15 September 1922), 4.

9. Pressure is, by definition, the force, or load, applied to each square inch of surface area. Mathematically, that is, Pressure = Force/Area. With pneumatic tires, increasing the force, or load, on the tire also increases the tire's contact area with the ground. The pressure exerted on the ground is therefore not a direct linear relationship to the load applied. Ronald Miller and David Sawers, *The Technical Development of Modern Aviation* (London: Routledge, 1968) is the classic work focusing on the achievement of greater efficiency, power, and performance in aircraft.

10. Bendix Landing Gear Service Bulletin, 1 March 1945, located in "Landing Gear, General," Technical File Y4000200, National Air and Space Museum Archives, Washington, DC.

11. Deborah Douglas, "Airports as Systems and Systems of Airports: Airports and Urban Development in America Before World War II," in Leary, *Airships to Airbus,* 55–86.

12. Nick A. Komons, *From Bonfires to Beacons: Federal Civil Aviation Policy under the Air Commerce Act, 1926–1938* (1978; reprint, Washington, DC: Smithsonian Institution Press, 1989), 130–31. This was the source of one of Foulois's early conflicts with Billy Mitchell,

who wanted airfields placed in the northeast, nearer to most of the army's infrastructure. See Foulois, *Wright Brothers to the Astronauts*, 139.

13. Eddie Rickenbaker referred to such days as "dud days" in his memoir of World War I flying (35).

14. Archibald Black, "Landing Field Roads and Runways," *Aeronautical Digest* (April 1923): 253–54, 292; (May 1923): 330–32.

15. Archibald Black, *Civil Airports and Airways* (New York: Simmons-Boardman Publishing Company, 1928).

16. Quincy Campbell, "Runways of Brick," *Airports* (September 1928): 8–9.

17. Douglas, "The Invention of Airports," 289–330.

18. C. N. Connor, "Airports and Transportation Engineering: Effect of Airplane Impact on Airport Surfaces," *Aviation Engineering* (December 1932): 18–21.

19. William F. Centner, letter to the editor, *Airports* (August 1929): 27, 28, 49; Wendell Miller, "The Drainage Factor in Airport Site Selection," *Airports* (March 1929): 9–10, 44.

20. The tendency of airport designers to minimize costs using centerline drains ended when a CAA study in the mid-1930s found that many runways had cracked after the pipes under them had collapsed. Airport engineers then adopted the current practice of putting the drainage piping down each edge of the runways, which at least doubled the amount of piping required; Edwin A. Miller, "Good Runways and Drainage at Rochester Airport," *Airports* (August 1929): 42–44, 58.

21. D. W. Crum, "Iowa City," *Airports* (March 1931): 19–20; "Airports in Pictures," *Airports* (July 1929): 32.

22. Jerold Brown, *Where Eagles Land* (New York: Greenwood Press, 1990, 40).

23. Ibid., 82.

24. Charles Stevenson, "Induction through Air and Water at Great Distances without the Use of Parallel Wires," *Proceeding of the Royal Society of Edinburgh*, 20 (1892–93): 25–27; H. Cooch, "Landing Aircraft in Fog," *Journal of the Royal Aeronautical Society* (June 1926): 365–93.

25. John S. Gray, "The Loth Leader Cable System for Electrical Steering of Aeroplanes," *Proceedings of the Institution of Aeronautical Engineers* 9 (1923): 7–30; Monte Duane Wright, *Most Probable Position: A History of Aerial Navigation to 1941* (Lawrence: University Press of Kansas, 1972); A. Verdurand and J. Blancard "Utilisation des procédés Loth pour le guidage des avions par ondes hertziennes," *L'Aérotechnique* (October 1930): 364–76; Paul Larivière, "Le câble de guidage-son emploi pour l'atterrissage sans visiblité," *L'Aéronautique* (March 1935): 33–39; Capitain P. Franck and Capitain A. Vomerange, "Le guidage des avions par câbles électriques," *L'Aéronautique* (1924): 39–47.

26. William Loth, "On the Problem of Guiding Aircraft in a Fog or by Night When There Is No Visibility," National Advisory Committee for Aeronautics Technical Memo 57, January 1932. Translated from *Comptes Rendus des Seances de l'Academie des Sciences*, no. 23, 5 December 1921.

27. A. Verdurand and J. Blancard, "Utilisation des procédés Loth pour le guidage des avions par ondes hertziennes," *L'Aérotechnique* (October 1930): 364–76; Paul Larivière, "Le câble de guidage-son emploi pour l'atterrissage sans visiblité," *L'Aéronautique* (March 1935):

33–39. Captain P. Franck and Captain A. Vomerange, "Le guidage des avions par câbles électriques," *L'Aéronautique* (1924): 39–47.

28. Charles Çhristienne and Pierre Lissarague, *A History of French Military Aviation,* trans. Francis Kianka (Washington, DC: Smithsonian Institution Press, 1986), 210–43, 310; Verdurand and Blancard, "Utilisation," 375.

29. Cooch, "Landing Aircraft in Fog," 365–93.

30. Ernest T. Williams criticized Cooch's paper, ibid., 388. However, Cooch discounted the comments (390–91) as did Mr. Handley Page (383).

31. Ibid., 391.

32. Ibid., 391; F. W. Meredith, "Air Transport in Fog," *Journal of the Royal Aeronautical Society* (February 1931): 75–85.

33. For a detailed discussion, see Chapter 3.

34. Tymms in Meredith, "Air Transport in Fog," 90. Emphasis added.

35. William M. Leary, "The Search for an Instrument Landing System, 1918–1948," in *Innovation and the Development of Flight,* ed. Roger D. Launius (College Station: Texas A&M University Press, 1999), 80–99.

36. Frederic Celler, "Landing Blind: The Loth System of Energized Cables for Fog Landing Guidance," *Aviation* (December 1931): 699–700; M. Heinrich Gloeckener, "Methods for Facilitating the Blind Landing of Airplanes," NACA Technical Memo 687, trans. by Dwight M. Miner from "Verfahren zur Erleichterung von Blindlandungen," *Zeitschrift für Flugtechnik und Motorluftschiffahrt* 23 (24 June 1932): 12.

37. "Hanson Announces Fog-Landing Device," *Aviation* (22 February 1930): 402.

38. Edward Nelson Dingley Jr., "An Instrument Landing System," *Communications* 18, no. 6 (June 1938): 7–9.

39. G. H. Mills to Chief of the Bureau of Ships, 14 February 1941; Chief of the Bureau of Ships to Chief of Naval Operations, 1 August 1941, both in file: C-F42-1/88, box 18, Bureau of Ships, General Correspondence 1940–1945, RG 19, NA. Stark's order is contained in the fourth indorsement to the basic letter.

1. A. Hunter Dupree, *Science in the Federal Government* (Baltimore: Johns Hopkins University Press, 1986), 271–77. The name National Bureau of Standards was in use from 1901 to 1903, when the organization became Bureau of Standards. That name lasted until 1934, when the organization was renamed National Bureau of Standards again. It is now the National Institute of Standards and Technology. For clarity, I will use Bureau of Standards throughout and abbreviate it NBS. William F. Snyder and Charles L. Bragaw, *Achievement in Radio* (Washington, DC: USGPO, 1986), 43.

2. The official history is Nick Komons, *Bonfires to Beacons: Federal Civil Aviation Policy under the Air Commerce Act, 1926–1938* (Washington, DC: Smithsonian Institution Press, 1989).

3. Ibid., 147–63.

4. H. Diamond, "Use of the Radio Beacon System for Landing in Fog," 19 November

1928, unpublished ms. in box 10, Papers of J. H. Dellinger, RG 167, NA. Diamond was born in Minsk, Russia, in 1900 and was naturalized a citizen on 4 June 1923. He had previously served as an instructor in electrical engineering at Lehigh University until joining the NBS in July 1927; the term "hertz" was not in use at the time, but for purposes of clarity, I will use it exclusively in this chapter. The historically correct term is "cycles per second," with the usual metric system prefixes "kilo-," "mega-," etc. One hertz is by definition one cycle per second, and they are therefore identical mathematically, if not historically.

5. Richard Hallion, *Legacy of Flight* (Seattle: University of Washington Press, 1977).

6. James H. Doolittle, "Flying an Airplane in Fog," *SAE Journal* (March 1930): 318–20, 345. Doolittle reported that the aircraft could withstand landing at 1,000 fps but he could not.

7. Doolittle's equipment is described in Hallion, *Legacy of Flight*, and The Daniel Guggenheim Fund for the Promotion of Aeronautics, *Equipment Used in Experiments to Solve the Problem of Fog Flying* (New York: The Guggenheim Fund, 1930). The airline estimate is United's. See J. R. Cunningham to Major Chester Snow, 16 November 1934, in file 827.1 vol. 1: July 1934–July 1937, folder 1 of 2, box 359, CAA Central Files, RG 237, NA.

8. E. F. W. Alexanderson, "Height of Airplane above Ground by Radio Echo," *Aviation Engineering* (December 1928): 12–13; A. P. Rowe, "Flying and Landing in Fog," *Aircraft Engineering* (July 1939): 169–72.

9. H. Diamond and F. W. Dunmore, "A Radio Beacon and Receiving System for Blind Landing of Aircraft," *Bureau of Standards Journal of Research, RP 238* (1930): 897–931; R. Schultz, "'Nebellanden' mit hilfe von Richtfunkbaken" [Fog Landing with Help from Directive Radio Beacons], *Zeitschrift für das Weltflugwesen* (March 1931): 129–34; Secretary of the Navy to Secretary of Commerce, 9 August 1933, file: F42-1/88 vol. 1, box 789, Bureau of Aeronautics Confidential Correspondence, RG 72, NA; an example is K. Baumann and A. Ettinger, "A New System for Blind Landing of Airplanes," *Proceedings of the Institute of Radio Engineers* 24, no. 5 (May 1936): 751–54; a long-running program to develop a system using very low frequency electromagnetic fields is detailed in the previous chapter.

10. Marshall Boggs to Col. H. H. Blee, 14 September 1931, file 827.1 vol. 1: June 1930–June 1934, folder 2 of 2, box 359, CAA Central Files, RG 237, NA.

11. Deborah G. Douglas, "The Invention of Airports" (Ph.D. diss., University of Pennsylvania, 1996), 248–50; W. S. Hinman Jr., "Preliminary Report on Tests of the Blind Landing Systems at Newark Airport," 23 May 1934, file: Blind Landing Newark 1932–1934, box 14, Papers of J. H. Dellinger, RG 167, NA.

12. Some controversy exists over the status of the Newark and Oakland installations. A 1939 *Aero Digest* article states that United got custody of the discarded Newark installation from the Bureau of Air Commerce in 1934 and transported it to Oakland. See Henry W. Roberts, "United Air Lines Radio Laboratory," *Aero Digest* (November 1939): 60, 62. Leary and Douglas have followed this article. Internal documents of the Bureau of Air Commerce, file 827.1 vol. 1: Inst. Landing Systems, box 359, CAA Central Files, RG 237, NA, and W. E. Jackson, "The Status of Instrument Landing Systems," CAA Technical Development Report no. 1 (Washington, DC: Civil Aeronautics Authority, October 1937) make clear that these were two separate installations. The Newark equipment was moved to Indianapolis in 1936, where the Bureau of Air Commerce began working on it again.

13. W. S. Hinsman Jr., "Preliminary Report on Tests of the Blind Landing Systems at Newark Airport," 23 May 1934, file: Blind Landing Newark 1932–1934, box 14, Papers of J. H. Dellinger, RG 167, NA.

14. Memo from G. Fulton to Head of Material Branch, Bureau of Aeronautics, 16 December 1935, file: F42-1/88 vol. 1, box 789, Bureau of Aeronautics Confidential Correspondence, RG 72, NA. The Washington Institute of Technology consisted of a number of laid-off Bureau of Standards radio researchers who were interested in continuing the blind landing work.

15. William Leary, "The Quest for an Instrument Landing System," *Innovation and the Technology of Flight,* ed. Roger D. Launius (State College, TX, 1999), 93.

16. Ibid., 80–99; Capt. George V. Holloman, "Inspection of Instrument Landing Systems," 24 June 1936, file 413.44, July–December 1936, box 1256, Office of the Chief Signal Officer Correspondence, RG 111, NA.

17. S. H. Ingersoll to Bureau of Engineering, 31 March 1939; Commander Aircraft, Scouting Force to Chief of the Bureau of Engineering, 13 December 1937; Capt. Mark Mitscher to Chief of the Bureau of Engineering, 11 August 1939, all in file: F42-1/88 vol. 1, box 789, Bureau of Aeronautics Confidential Correspondence, RG 72, NA.

18. Roy Jackson to Chief of the Bureau of Engineering, 31 May 1940, file "F-42-1/88" vol. 2, box 789, Bureau of Aeronautics Confidential Correspondence, RG 72, NA.

19. Commander, Patrol Wing Seven to Chief of the Bureau of Aeronautics, 14 February 1942, file "F-42-1/88" vol. 3, box 789; Captain C. A. Nicholson to the Chief of the Bureau of Aeronautics, 10 November 1943, and Chief of BuAer to Chief of BuShips, 10 October 1943, file "F42-9," box 892, all in Bureau of Aeronautics Confidential Correspondence, RG 72, NA.

20. Foulois, *From the Wright Brothers to the Astronauts* (New York: McGraw-Hill, 1968), 207, 224.

21. Monte Duane Wright, *Most Probable Position: A History of Aerial Navigation to 1941* (Lawrence: University Press of Kansas, 1972), 86–87, 90–94.

22. Ibid., 127.

23. Foulois claims in his memoirs that the army's work of 1934 was the beginning of ILS and GCA; *Wright Brothers to the Astronauts,* 258; Chester Snow to Rex Martin, 6 June 1934, file 827.1 vol. 1: June 1930–June 1934, folder 2 of 2, box 359, CAA Central Files, RG 237, NA.

24. Cunningham to Eugene Vidal, 31 October 1934; C. C. Shangraw, 28 November 1934; Walter J. Addems, 5 November 1934; H. M. Hucke to Cunningham, 7 November 1934; H. B. Sneed to Jack Frye, 4 January 1934, all extracted in Lloyd L. Juleson to Rex Martin, 5 April 1935, file 827.1: July 1934–July 193,, folder 1 of 2, box 359, CAA Central Files, RG 237, NA.

25. C. C. Shangraw, 28 November 1934, ibid.

26. J. R. Cunningham to J. Lyman Briggs, 1 May 1935; Chief of the Bureau of Aeronautics to Chief of Naval Operations, 25 November 1935, both in file "F42-1/88," box 7, Bureau of Engineering Confidential Correspondence 1940–1945, RG 19, NA.

27. W. E. Jackson, "The Status of Instrument Landing Systems," *Technical Development Report no. 1* (Washington, DC: Civil Aeronautics Authority, October 1937).

28. I have not been able to find reliable ceiling statistics for the 1930s. A 1949 document puts the percent of time below a 500-foot minimum for La Guardia at 5.5 percent, and for Los Angeles, 5.2 percent. See Edgar A. Post, "Airline Operating Experiences with the Instrument Landing System," Institute of Aeronautical Sciences, 22 July 1949.

29. Diamond and Dunmore, "Radio Beacon and Receiving System"; R. Stüssel, "The Problem of Landing Commercial Aircraft in Fog," *Journal of the Royal Aeronautical Society* (1934): 807–36. The Lorenz system was produced in conjunction with the German aeronautical research institute, Deutsche Versuchsanstalt für Luftfahrt; Jackson, "Status of Instrument Landing Systems," 7–9.

30. Carl Lorentz A. G. to Major Gardner, 18 March 1936; First indorsement to Lt. Col. C. K. Nulsen to Military Attache, Berlin, Germany, 27 January 1936, both in folder 413.44: Inst. Landing no. 1, box 1256, Office of the Chief Signal Officer Correspondence, RG 111, NA.

31. First indorsement to Lt. Col. H. H. Fuller to Assistant Chief of Staff (G-2), 19 February 1936; third indorsement to Lt. Col. H. H. Fuller, ibid.

32. Wright, *Most Probable Position*, 103–29.

CHAPTER 4: THE PROMISE OF MICROWAVES

1. William M. Leary, "The Search for an Instrument Landing System, 1918–1948," in *Innovation and the Development of Flight*, ed. Roger Launius (College Station: Texas A&M University Press, 1999), 94–95.

2. William Osmun, *The Authority of Agreement: A History of RTCA* (Washington, DC: RTCA, 1985).

3. Captain G. V. Holloman and Major F. S. Borum, "Lorenz Instrument Landing System," file 413.44: Inst. Landing no. 4, box 1256, Office of the Chief Signal Officer Correspondence, RG 111, NA.

4. John O. Mauborgne to Chief Signal Officer, 19 May 1937, ibid. The Chief Signal Officer of the Army was responsible for radio development within the Army, including aircraft radio; therefore the Director of the Aircraft Radio Laboratory at Wright Field worked for him. This required considerable coordination with the Army Air Corps.

5. Noise abatement requirements ended the "straight in approach" during the 1970s, except in bad weather. The ILS, obviously, permits only straight approaches.

6. "Pilots Ultimatum Condemns Washington Airport"; "Fears for Friends Flying to Capital, Mrs. Roosevelt Says," both in *The Air Line Pilot* (July 1937): 1; "An Editorial in Pictures," *The Air Line Pilot* (November 1937): 1. The editorial was reprinted from the *Washington Times*; "Ask Removal of Hazardous Obstructions," *The Air Line Pilot* (November 1937): 4; FDR's decision was to move the airport to Gravelly Point on the Virginia bank of the Potomac River; it later became National Airport.

7. W. E. Jackson to Gazely, 28 June 1937, file 827.1: July 1934–July 1937, folder 1 of 2, box 359, CAA Central Files, RG 237, NA; David Behncke, "New Developments on Instrument Landing System," *The Air Line Pilot* 6 (August 1937): 1, 5, 7, quote p. 5. Freng was Superintendent of Flying for United at this time; Paul Goldsborough to Major A. W. Marriner, 25 June 1937, file 413.44: Inst. Landing no. 4, box 1257, Office of the Chief Signal Officer

Correspondence, RG III, NA. Marriner communicated extensively with the Aircraft Radio Laboratory, which probably explains why a copy of this letter is in the Signal Office's files. Donald MacKenzie discusses the influence of ARINC in the context of the laser ring gyroscope in "From the Luminiferous Ether to the Boeing 757: A History of the Laser Ring Gyroscope," *Technology and Culture* (1993): 475–515.

8. Secretary of War to Secretary of Commerce, 21 April 1938; Col. Frank M. Kennedy to the Chief of the Air Corps, 13 April 1938, both in file 413.44: Inst. Landing no. 5, box 1257, Office of the Chief Signal Officer Correspondence, RG III, NA.

9. H. H. Buttner and A. G. Kandorian, "Development of Aircraft Instrument Landing Systems," *Electrical Communications* 22, no. 2 (1945): 179–92.

10. Report of Meeting of Subcommittee #4, Instrument Landing Devices, Radio Technical Committee for Aeronautics, file 827.1 vol. 2: November 1938–December 1939, folder 1 of 2, box 359, CAA Central Files, RG 237, NA. For the record, the pilots were Otis Bryan, TWA; Fred Davis, Eastern; R. T. Freng, United; E. A. Cuttrell and B. O. Howard, American. Northwest Air Lines and National Air Line were also represented, but not, apparently, by a pilot.

11. Ibid., 2.

12. Preston Bassett, 27 October 1939, file: Instrument Landing, box 35, Sperry Gyroscope Company Collection, Hagley Museum and Library, Wilmington, Delaware.

13. Report of Meeting of Subcommittee #4.

14. Ibid., 14. Arnold later concurred with Mitchell's assessment in an independent memo to Clinton Hester, 2 November 1939, file 413.44, box 673, AAF Central Decimal Files, RG 18, NA.

15. Report of Meeting of Subcommittee #4, 8.

16. Ibid.

17. John M. R. Wilson, *Turbulence Aloft: The Civil Aeronautics Administration in War and Rumors of War, 1938–1958* (Washington: FAA, 1980), 1–42; CAA had no authority over military aviation, unlike the current FAA. That problem was not rectified until 1958, when CAA (which was organizationally revised in 1941 into an Administration, operating under the Commerce Department) was replaced with the FAA.

18. See W. L. Barrow and L. J. Chu, "Theory of the Electromagnetic Horn," *Journal of the Institute of Radio Engineers* 27, no. 1 (January 1939): 51.

19. Donald G. Fink, "3 Spots and a Horn," *Aviation* 37 (September 1938): 28, 29, 73, 74; Alex Soojung-Kim Pang, "Edward Bowles and Radio Engineering at MIT, 1920–1940," *Historical Studies of the Physical Sciences* 20, vol. 2 (1990): 313–37.

20. E. L. Bowles, W. L. Barrow, W. M. Hall, F. D. Lewis, and D. E. Kerr, "The CAA-MIT Microwave Instrument Landing System," preprint of AIEE paper dated 2 December 1939, in file 413.33: Inst. Landing no. 6, box 1458, Office of the Chief Signal Officer Central Files, RG III, NA.

21. Arthur L. Norberg and Robert W. Seidel, "The Contexts for the Development of Radar: A Comparison of Efforts in the United States and the United Kingdom in the 1930s," *Tracking the History of Radar,* ed. Oskar Blumtritt, Hartmut Petzold, and William Aspray (Piscataway, NJ: IEEE, 1994), 199–216; undated ts., folder 8, box 4, Russell Varian series, Varian Papers, SC 345, Stanford University Archives.

22. Undated ts., Russell Varian series; copy of agreement, 26 April 1938, folder 11, box 5, ibid. This later agreement references the original dated 6 October 1937.

23. Undated ts.; agreement between Leland Stanford Junior University and Sperry Gyroscope Company, 27 April 1938, folder 11, box 5, Russell Varian series.

24. The Chief Signal Officer was a Major General billet, and upon promotion to Brigadier General and assignment as CSO, he was breveted to Major General. The visit appears to have been arranged by the Office of the President, in the White House. Compton transmitted his thanks for the visit through a memo addressed to the Office of the President, which then wrote to Mauborgne. See Allen W. Horton to Maj. Gen. John Mauborgne, 30 April 1938, file 413.44, box 1256, Office of the Chief Signal Officer Correspondence, RG 111, NA.

25. Col. Hugh Mitchell to Col. Bender (the addressee is "My Dear Bender"), 26 March 1938, ibid.

26. Sperry Gyroscope Company, "A Proposal for an Instrument Landing System," n.d. [late 1939], ts., file: Instrument Landing #21, box 35, Sperry Gyroscope Company Collection, Hagley Museum and Library.

27. Clinton Hester to Maj. Gen. H. H. Arnold, file 413.44, box 673, AAF Central Decimal Files, RG 18, NA.

28. Maj. Gen. H. H. Arnold to Assistant Secretary of War, 20 August 1939, ibid.; H. H. Arnold, *Global Mission* (New York: Arno Press, 1949), 165.

29. Franklin D. Roosevelt to Dr. Paul Brockett, 30 August 1939; Frank B. Jewett to President of the United States, 1 September 1939, both in file 413.44, box 673, AAF Central Decimal Files, RG 18, NA.

30. Frank Jewett to President, 24 November 1939, ibid.

31. Charles Stanton to Vannevar Bush, 14 October 1939, file 827.1 vol. 2: November 1938–December 1939, folder 1 of 2, box 359, CAA Central Files, RG 237, NA.

32. Edgar S. Gorrell to Vannevar Bush, 30 September 1939, Committee on Airplane Instrument Landing Equipment Collection, National Academy of Sciences Archives, Washington, DC (NAS Archives).

33. Bush, head of the Carnegie Institute of Washington, was also a member of MIT's Electrical Engineering faculty and therefore not quite a disinterested party.

34. Lt. Col. Hugh Mitchell to Richard Gazely, 25 October 1939, General file, Committee on Airplane Instrument Landing Equipment Collection, NAS Archives. Mitchell's statement is a summation of one sent by Maj. Gen. Arnold to Clinton Hester, 2 November 1939, which also stated that the straight glide path should extend to at least ten miles. Letter in file 413.44, box 673, AAF Central Decimal Files, RG 18, NA.

35. Edward L. Bowles to Vannevar Bush, 2 November 1939, Committee on Airplane Instrument Landing Equipment Collection, NAS Archives.

36. Ibid.

37. He referred to ultra-high frequency, but he clearly meant what we call (and others at the time called) microwave. At this time, the frequency spectrum did not yet have its current structure, and different terms were in use for the same frequency band. Col. Mitchell, for example, referred to Bowles' work with super-high frequencies, a term Bowles never used.

38. Bowles to Bush, 2 November 1939.

39. Ibid.

40. Vannevar Bush to Members of the Conference Committee on Instrument Landing Systems, 20 November 1939, General file, Committee on Airplane Instrument Landing Equipment Collection, NAS Archives.

41. Ibid.

42. Vannevar Bush to Frank Jewett, 21 November 1939; Frank Jewett to President Roosevelt, 24 November 1939, file 413.44B; Brig. Gen. Edwin Watson to Hon. Harry H. Woodring, 5 December 1939, cited in Harry H. Woodring to Brig. Gen. Edwin Watson, 28 December 1939, file 413.44, all in box 673, AAF Central Decimal Files, RG 18, NA. Clinton M. Hester to Vannevar Bush, 18 December 1939, General file, Committee on Airplane Instrument Landing Equipment Collection, NAS Archives.

43. Paul Brockett to Frank Jewett, 21 December 21; Paul Brockett to Frank Jewett, 26 December 1939, both in General file, Committee on Airplane Instrument Landing Equipment Collection, NAS Archives. It isn't clear whether the White House was irritated with the Authority, a five-member board that directed the agency, or the Administrator, who was supposed to actually run it. The confusing structure of the CAA was resolved in 1940, when the Authority was separated out as the Civil Aeronautics Board, and the administrative functions were placed in the Civil Aeronautics Administration, a unit of the Commerce Department.

44. Woodring to Watson, 28 December 1939. Woodring's reply was taken verbatim from Maj. Gen. Arnold's response to the report. See 2nd indorsement to Adjutant General to Chief of the Air Corps, 20 December 1939, file 334.8: National Academy of Sciences, box 1773, Office of the Adjutant General Central Files, RG 407, NA.

45. Maj. Gen. H. H. Arnold to Chief Signal Officer, 20 December 1939, file 413.44, box 673, AAF Central Decimal Files, RG 18, NA.

46. Franklin Roosevelt to Brig. Gen. Edwin Watson, 5 January 1940, ibid.; Adjutant General to Chief Signal Officer and Chief of the Air Corps, 9 January 1940, file 334.8: National Academy of Sciences, box 1773, Office of the Adjutant General Central Files, RG 407, NA.

47. Second indorsement to Adjutant General to Chief Signal Officer and Chief of the Air Corps, 11 January 1940; Harry Woodring to Brig. Gen. Watson, 16 January 1940, file 334.8: National Academy of Sciences, box 1773, Office of the Adjutant General Central Files, RG 407, NA.

48. Maj. Gen. Arnold to Brig. Gen. Watson, 12 February 1940, folder 413.44, box 673, AAF Central Decimal Files, RG 18.

49. Clinton Hester to Maj. Gen. H. H. Arnold, 12 April 1939; Sixth indorsement to Aircraft Radio Laboratory to Chief Signal Officer, 10 October 1939, ibid. The Sixth indorsement is signed by Arnold, under date 20 December 1939. Second indorsement to Adjutant General to Chief of the Air Corps, 20 December 1939, file 334.8: National Academy of Sciences, box 1773, Office of the Adjutant General Central Files, RG 407, NA. Note that the date on this indorsement is the same as the date under which Arnold asked the Chief Signal Officer to contract with Sperry for the microwave system. Given that the Chief Signal Officer's memo had begun its trek through the bureaucracy in early October, it seems very

likely that Arnold was waiting for the NAS recommendations before making his own recommendation to the CSO. In his response, Arnold quoted the committee's microwave recommendations verbatim.

50. Anonymous to Maj. General Arnold, 12 March 1940, file 413.44, box 673, AAF Central Decimal Files, RG 18, NA.

51. Arnold, *Global Mission*, 165; Wilson, *Turbulence Aloft*, 28.

52. Lt. Col. Orlando Ward to the Chief of the Air Corps, 3 April 1940; Brig. Gen. R. K Yount to the Secretary of War, 13 April 1940; draft of memo (with routing info) in Office of the Adjutant General, all in file 334.8: National Academy of Sciences, box 1773, Office of the Adjutant General, RG 407, NA. Final and "EMW" (Brig. Gen. Edwin Watson) to The President, 2 May 1940, in file 413.4B, box 673, AAF Central Files, RG 18, NA.

53. Pan American, of course, used seaplanes extensively, but that organization had its own system. Since it was not technically a domestic airline, it also does not seem to have been invited to these conferences. It certainly never attended.

CHAPTER 5: INSTRUMENT LANDING GOES TO WAR

1. The glide path was 330 MHz. This was a UHF frequency until 1944 when it was redesignated VHF.

2. Major E. M. Powers, 20 June 1940, file 413.44, box 673, AAF Central Decimal Files, RG 18, NA.

3. According to the Army, seven sets built with 1940 and 1941 money and installed during 1941 were Chicago, Cleveland, Fort Worth, Kansas, Los Angeles, New York/La Guardia, and Washington, DC. Office of the Chief of the Army Air Forces, "Air Force Communications Policy: Instrument Approach and Landing Systems," 31 October 1941, in unmarked folder, box 404, RD 158, Records of Wright Field (Sarah Clarke Collection), RG 342, NA.

4. Sir Charles Webster and Noble Frankland, *The Strategic Air Offensive Against Germany, 1939–1945* (London: Her Majesty's Stationary Office, 1961); H. H. Arnold, *Global Mission* (New York: Arno Press, 1949), 215–40; David B. Langmuir to Lee DuBridge, 14 January 1943, file: Project 102, box 404, RD 158, Records of Wright Field (Sarah Clarke Collection), RG 342, NA.

5. See "Air Force Communications Policy," n. 186; David Little, 21 November 1941, file 413.44: Inst. Landing no. 7, box 1459, Office of the Chief Signal Officer Central Files, RG 111, NA.

6. Henry L. Stimson to Secretary of Commerce, 19 May 1942, file 413.44: Inst. Landing no. 8, box 1459, Office of the Chief Signal Officer Central Files, RG 111, NA. Although the date of the official letter is May, the arrangements for this program began in January.

7. International Telephone Development Corp., a subsidiary of International Telephone and Telegraph, changed its name to International Telephone and Radio Manufacturing Company (ITRM), in 1941, and then to Federal Telephone and Telegraph Company, apparently in 1943. The name changes were likely due to its connection to ITT, which was attacked in the press, and investigated by the military, over its subsidiaries in Germany (of which Lorenz A.G. was one).

8. John M. R. Wilson, *Turbulence Aloft: The Civil Aeronautics Administration during Wars and Rumors of War* (Washington, DC: FAA, 1979). The agency was the Interdepartmental Air Traffic Control Board.

9. See, for example, CNO to CG, AAF, 21 June 1944, file: F42-9 vol. 1, box 892, Bureau of Aeronautics Confidential Correspondence, RG 72, NA.

10. Major Lyman D. Swendson to Chief, Maintenance Division, Air Service Command, 11 May 1944, file: Project 102, box 404, RD 158, Records of Wright Field (Sarah Clarke Collection), RG 342, NA, contains the allocation and priority lists for the various theaters. Receivers were held up for months at shipping depots: see Deputy Directory, ATSC, to CG ETO, 14 September 1944, file: Project 102 (UK); Major F. L. Moseley, "Visit Report to Air Transport Command and Office of the Chief Signal Officer," 10 November 1943, file: Proj. 102 (CAA Correspondence), both in ibid. By May 1944, there were twelve in operation in the United States.

11. William Rose to author, 24 March 1998, personal communication. Rose was a pilot assigned to 92nd BG (H). The ceiling data is from Strategic Bombing Survey, *Weather Factors in Combat Bombardment Operations in the European Theater*, 2nd ed. (January 1947): 9A; J. W. Howland to author, 23 March 1998, personal communication; see also Rose to author, 24 March 1998. Apparently the Gee method also only worked at certain airfields. Howland served with 381st BG (H) and 91st BG (H).

12. Col. Daniel C. Doubleday to Chief Signal Officer, 8 November 1943; European Theater of Operations (Eisenhower) to Wright Field, telegram dated 8 February 1944; War (Secretary of War) to ASCPFO, 13 March 1944, all in file: Project 102 (UK), box 404, RD 158, Records of Wright Field (Sarah Clarke Collection), RG 342, NA.

13. Frank Griffiths, *Angel Visits: From Biplane to Jet* (London: Thomas Harnsworth, 1986): 101–103. Arnold to McClellan (signed by Spaatz), telegram, 14 March 1944, file: Project 102 (UK), box 404, RD 158; Carl Spaatz to [recipient unreadable, but probably General McClelland], 13 May 1944, file: Project 102D, box 405, RD 159; 1st Lt. Donald Hansen, 29 January 1945, file: Project 102 (UK), box 404, RD 158, all in Records of Wright Field (Sarah Clarke Collection), RG 342, NA.

14. Wilbert Swank, 8 May 1944, file: Project 102; 1st Lt. Donald Hansen, 29 January 1945, file: Project 102 (UK), both in box 404, RD 158, Records of Wright Field (Sarah Clarke Collection), RG 342, NA. This author noted that SBA's failure continued to cause hesitation to use SCS-51.

15. HQ USSTAF, London, to ATSCWFO, telegram dated 12 September 1944, file: Project 102 (UK), box 404, RD 158, Records of Wright Field (Sarah Clarke Collection), RG 342, NA.

16. See Director of Bombardment to Director of Communications, 8 January 1943, box 1107, AAF General Correspondence, RG 18, NA.

17. Hansen, 29 January 1945 (see n. 14). See also Robert B. Parke, *The Pilot Maker* (New York: Grosset and Dunlap, 1970), on the Link Trainer.

18. Only twenty-five were installed. The remaining five were dispatched to the continent after the Normandy invasion, where they were used to mark the front lines so that pilots would know when they were over friendly territory. This was done to reduce the prob-

ability of aircraft mistakenly attacking friendly ground forces. With the localizer aligned parallel to the front, the 90 Hz signal zone marked friendly territory, and the 150 Hz zone marked enemy territory. Telephone conversation with Timothy Bland, 23 March 1998.

19. Preston R. Basset to T. A. Morgan, 18 July 1941, file: Instrument Landing, box 35, Sperry Gyroscope Collection, Hagley Museum and Library.

20. Hugh G. J. Aitken, *Continuous Wave: Technology and American Radio, 1900–1932* (Princeton: Princeton University Press, 1985), 1–27. The term "intellectual tradition" is my own formulation of his much more complex argument. See also his earlier work on radiotelegraphy: Hugh G. J. Aitken, *Syntony and Spark: The Origins of Radio* (Princeton: Princeton University Press, 1976).

21. More detail on the Lab's founding can be found in Chapter 6.

22. A. L. Loomis, "Report of the Ad Hoc Committee on Instrument Landing," 16 February 1942, box 53A, Records of the MIT Radiation Lab, RG 227, NA.

23. J. H. Buck to Lee DuBridge, 2 December 1942, file: PGP [1941–1943], box 54A, MIT Radiation Laboratory Records, RG 227, NA; for comments suggesting this interpretation, see Buck to DuBridge; Robert Davies to J. H. Buck, 13 April 1943, ibid.; Loomis, "Report"; MIT Radiation Laboratory Records, RG 227, NA.

24. Buck to DuBridge, 2 December 1942.

25. Ibid.; Henry Guerlac, *History of Radar in World War II* (New York: Tomash Publishers, 1987).

26. The best work on ICAO's foundation is Alan Dobson, *Peaceful Air Warfare: The United States, Britain, and the Politics of International Aviation* (Cambridge: Oxford University Press, 1991).

27. On Watson-Watt, see his autobiography: Robert Watson-Watt, *Three Steps to Victory* (London: Odhams, 1957); on Stanton, see Wilson, *Turbulence Aloft*, 137–40.

28. Robert Watson-Watt, "Record of an Informal Fourth CERCA meeting held on Sunday, 10 November 1946," COT/15, PICAO documents, International Civil Aviation Organization (ICAO) Library and Archives, Montreal, Canada. CERCA stood for "Commonwealth and Empire Radio Committee Assembly."

29. Special Radio Technical Division, "Minutes of the Tenth Meeting," 11 November 1946, COT/21, PICAO documents, ICAO. Stanton's criticism of Watson-Watt is in COT/22, same date.

30. MATS had stripped a number of lesser-used airways in the United States of their ranges during the war and shipped them overseas, where they were installed to facilitate cargo and aircraft movement. See Wilson, *Turbulence Aloft*, 119.

31. Special Radio Technical Division, "Minutes of First Meeting," 30 October 1946, PICAO documents, vol. 23, ICAO Library and Archives, Montreal.

32. Edward P. Warner, "Work of the Interim Council of the PICAO," *Aero Digest* (March 1946): 24–25, 148.

CHAPTER 6: THE INTRUSION OF NEWCOMERS

1. Robert Buderi, *The Invention That Changed the World: How a Small Group of Radar Pioneers Won the Second World War and Launched a Technological Revolution* (New York: Si-

mon & Schuster, 1996), 38–51; James Phinney Baxter, *Scientists against Time* (Cambridge, MA: MIT Press, 1946), 3–25.

2. Buderi, *Invention That Changed the World*, 38–51.

3. J. L. Heilbron and Robert W. Seidel, *Lawrence and His Laboratory: A History of the Lawrence Berkeley Laboratory*, Vol. 1 (Berkeley: University of California Press, 1989), 493.

4. Ibid.; Buderi, *Invention That Changed the World*, 46–49; Luis W. Alvarez, *Adventures of a Physicist* (New York: Basic Books, 1987), 88; Lawrence Johnston, interview with Frederik Nebeker, 13 June 1991, transcript, MIT Radiation Laboratory Oral History Collection, IEEE Center for Electrical History, Rutgers, NJ.

5. Alvarez, *Adventures*, 32–49; Heilbron and Seidel, *Lawrence and His Laboratory*, 236–37.

6. Alvarez, *Adventures*, 31.

7. Johnston interview; Buderi, *Invention That Changed the World*, 43.

8. The name was originally Ground Controlled Landing (GCL), but the Army insisted that blind landings were impossible, drawing Alvarez's scorn. The name was changed nonetheless, and of course the Army was proven correct. Although GCA could land aircraft completely blind sometimes, it could not do so with perfect reliability. That has been true of every blind landing system. At least two other approach systems based on the ground control model were invented during the war, both by technicians in combat theaters before Alvarez's GCA was deployed. In both cases, radars designed for airborne use (H2S/APS-15 in one case, and SCR-717 in the other) were modified and set up at an airfield to scan the runway approach sector. An operator then gave pilots landing directions via voice radio. George Reynolds to author, 23 March 1998, personal communication; "Baby GCA [1944–1945]," box 35B, MIT Radiation Laboratory Records, RG 227, NA.

9. The LSE still has a job, and modern carrier pilots also have a stabilized visual reference system on the ship's deck to assist in the approach. Certain aircraft can utilize a fully automatic radar-based "carrier controlled approach" system developed between the mid-1950s and the late 1960s.

10. A. L. Loomis, "Report of the Ad Hoc Committee on Instrument Landing," 16 February 1942, file: Report of Ad Hoc Committee on Instrument Landing, box 53A, Records of the MIT Radiation Lab, RG 227, NA.

11. Alvarez, *Adventures*, 86–110, and Buderi, *Invention That Changed the World*, 137–38.

12. Lawrence Johnston, "GCA: Ground Controlled Approach, Radiation Laboratory Report 438" (Cambridge: MIT Radiation Laboratory, 1 October 1943, photocopy of copy #87), file: RB 334.8 NDRC RL Report ET-2047, box 1428, Office of the Chief Signal Officer Central Files, RG 111, NA.

13. Alvarez, *Adventures*, 98. See also Jennet Conant, *Tuxedo Park: A Wall Street Tycoon and the Secret Palace of Science That Changed the Course of World War II* (New York: Simon & Schuster, 2002), 260.

14. Homer Tasker, "Technical Report of Radio Set AN/MPN-1 (XE-1)," Report no. 103, 15 June 1945 (Los Angeles: Gilfillan Bros., Inc., photocopy) file: GCA[GCL] GCA-Mk II [AN/MPN-1] GCA Mk III [1941–1945], box 53B, MIT Radiation Laboratory Records, RG 227, NA.

15. Alvarez, *Adventures*, 99–100; the military designation for the Gilfillan sets was AN/MPN-1(A).

16. Luis Alvarez, "Minutes of the Coordinating Committee," 20 January 1943, file: "Group 73, Landing [1942–1943]," box 41B, MIT Radiation Laboratory Records, RG 227, NA; Alvarez, *Adventures*, 100; Johnston, interview.

17. Johnston, interview. Date from Johnston, "GCA," 46 (see n. 12); Commander Fleet Air, Quonset Point, to Chief of the Bureau of Aeronautics, 26 April 1943, file: F42-9, box 892, Bureau of Aeronautics Records, RG 72, NA.

18. Tasker, "Technical Report," 8–9. The Mk I required 4,400 drawings totaling 12,800 square feet.

19. A fascinating incident happened during this test series when two RAF Mosquito bombers, one painted silver, the other black, at the same altitude, gave two very different readings. The black one disappeared from the scope at a distance of ten miles while broadside to the antenna; the silver one stayed visible until twenty miles out. The AAF took considerable notice: Lt. Lawrence McFadden to Chief Signal Officer, 25 August 1943, file: GCA Correspondence 1943, box 53, MIT Radiation Laboratory Records, RG 227, NA; the information in this paragraph has been synthesized from several sources: Johnston interview; Alvarez, *Adventures;* Clarke, *Glide Path;* George Comstock, transcript of interview with Henry Guerlac, 7 December 1944, in file: GCA/GCL Mk II [AN/MPN-1] Mk III [1941–1945], box 53B, MIT Radiation Laboratory Records, RG 227, NA. Although Clarke's work is a novel, enough of the events it reports match with those reported in the other sources to consider it to be a largely true account, even if it differs in its chronology.

20. Anonymous memo to the Coordinating Committee, 22 May 1945, file: GCA Memoranda, box 65A, MIT Radiation Laboratory Records, RG 227, NA.

21. Saipan and Iwo Jima. On Iwo Jima, P-61s with experienced pilots could be brought in under ceilings down to 25 feet, but the base implemented a policy of ordering crews to bail out over the island (directed by the GCA operator) for ceilings of under 100 feet because pilots who could not see below that point, even if landed safely, tended to collide with ground obstacles such as parked aircraft. Anonymous, undated document [c. June 1945], file: Baby GCA, box 35A, MIT Radiation Laboratory Records, RG 227, NA.

22. Frank Griffiths, *Angel Visits: From Biplane to Jet* (London: Thomas Harnsworth Publishing, 1986), 99.

CHAPTER 7: THE POLITICS OF BLIND LANDING

1. *AOPA Pilot* was a members-only insert to *Flying* magazine, provided by AOPA's staff. Because *Flying* had no editorial control over the insert, I have documented it as an independent publication. It is essentially a newsletter, and library editions of *Flying* did not contain it. It is difficult for scholars to access as a result, but the Hagley Museum and Library contains a complete set. *AOPA Pilot* became an independent publication in 1958.

2. Quote from Charles V. Murphy,"The Last 500 Feet," *Life* (12 May 1947): 93; Minutes of the Robert J. Collier Trophy Committee, 8 August 1946, Collier Trophy Collection, National Air and Space Museum Archives, Washington, DC; report prepared by Carl Hinshaw, *Aids to Air Navigation and Landing,* House Committee on Interstate and Foreign Commerce, Subcommittee on Air Safety, 80th Cong., 1st sess., 11 July 1947, 17–18. This

chapter appeared in a shortened form as Erik M. Conway, "The Politics of Blind Landing," *Technology and Culture* 42, no. 1 (January 2001): 81–106.

3. For clarity's sake, I will refer to both SCS-51 and ILS as ILS and both AN/MPN-1 and GCA as GCA. Although GCA is less a piece of equipment than a procedure, the distinction was not made at the time, and all references to GCA in the literature are to Alvarez's invention.

4. John M. R. Wilson, *Turbulence Aloft: The Civil Aeronautics Administration amid Wars and Rumors of Wars, 1938–1953* (Washington, DC: FAA, 1979), 164–68; Joseph Corn, *The Winged Gospel* (New York: Oxford University Press, 1983).

5. A parallel Senate investigation also took place. The House record is more detailed and complete, and I have therefore relied on it.

6. John Stuart, "Aviation Accidents: Ground Control Approach System, Backed by Many, Runs into Cost Objections," *New York Times* (2 February 1947): 14.

7. AAF Instrument Flying Standardization Board, "Recommendations of the Staff Study of the Relative Merits of SCS-51 and GCA," 26 May 1945, 202.2-54, attachment 225, Air Force Historical Research Agency (AFHRA), Maxwell AFB, AL; Air Transport Command, "Report on the Pilot and Instrument Approach System Evaluation Tests Conducted at Indianapolis," 11 July 1946, K239.0249-4609AU, AFHRA. The Air University's evaluation of the Indianapolis tests was requested by Gen. Carl Spaatz by memo dated 1 August 1946 and was forwarded to AAF Headquarters on 18 March 1947. Both are attachments to K239.0249-4609AU, AFHRA.

8. And, as if in a mirror, the pilots' union demanded radar control of the airways (but not radar-directed landing) during the 1950s, after a series of spectacular midair collisions demonstrated that the skies had become too crowded to permit the absolute freedom that pilots had traditionally enjoyed. They also fought to ensure that ground controllers' authority was carefully circumscribed, however. See Stuart I. Rochester, *Takeoff at Mid-Century: Federal Civil Aviation Policy in the Eisenhower Years, 1953–1961* (Washington, DC, Federal Aviation Authority, 1976), 57–78.

9. Navy pilots retained the right to ignore flight deck orders because the person issuing them was an enlisted rating, and by law enlisted ranks cannot give orders to officers—and almost all Navy pilots were officers. Hence, legally, the Navy had to allow pilots the right to ignore directions from the LSE, as the position is known. Further, the Navy's tradition of independence of command applied to aircraft as well as ships. The Navy therefore *expected* its pilots to accept directions from the flight deck, but it could not force them to. Navy pilots thus retained legal autonomy while still typically submitting to deck-based directions.

10. Commerce Member, IATCB to Director of Federal Airways, 10 February 1945, file: 504.0 vol. 6, box 83, CAA Central Files, RG 237, NA.

11. Chief, Ultra-High Frequency Unit to Acting Chief, Radio Engineering Section, 22 May 1945, ibid.

12. "Washington Observer—CAA Touchy on GCA," *Aviation News* (3 March 1947): 3.

13. Chief, Air Traffic Control Division, to Director of Federal Airways, 17 February 1945, file 504.0 vol. 6, box 83, CAA Central Files, RG 237, NA.

14. Ibid.; Thomas Bourne to Brig. Gen. H. M. McClelland, 14 February 1945, file 818.1

vol. 1, box 285, CAA Central Files, RG 237, NA. Unfortunately, Kroger's version is the one that CAA's official historian, Wilson, relied on. See William Kroger, "CAA and Airlines against GCA," *Air Transport* (March 1947): 30–33, 77–78; and Wilson, *Turbulence Aloft*, 323. I used the FAA History Office's annotated copy of Kroger's article in this analysis.

15. Wilson, *Turbulence Aloft*, 217; Ben Stern to Maurice Roddy, 12 November 1946, file 818.1: ASR vol. 1, box 825, CAA Central Files, RG 237, NA.

16. Gilbert later became one of the leading authorities on Air Traffic Control matters. See, for example, Glenn A. Gilbert, *Air Traffic Control: The Uncrowded Sky* (Washington, DC: Smithsonian Institute Press, 1973); Nick Komons, "Federal Government Helped Forge New Enterprise, New Profession," *Aviation's Indispensable Partner Turns 50*, Department of Transportation, n.d.

17. Chief, Air Traffic Control Division, to Director of Federal Airways, 17 February 1945, file 504.0 vol. 6, box 83, CAA Central Files, RG 237, NA. In combat theaters SCS-51s and runways were utilized at landing rates of up to one plane every thirty seconds. AN/MPN-1s were occasionally used at equivalent rates, but that seems to have been accomplished by landing several aircraft simultaneously, in formation. That was obviously only possible with small, fighter-sized aircraft. Such rates were clearly too risky for commercial aviation.

18. Ibid.

19. Asst. Chief, Airways Engineering Division, to Director of Federal Airways, 27 February 1945; Chief, Air Traffic Control Division, to Director of Federal Airways, 17 February 1945, both in file 504.0 vol. 6, box 84, CAA Central Files, RG 237, NA.

20. William Kroger, "Standardized System Expected in Instrument Landing Dispute," *Aviation News* (10 December 1945): 9–10; Robert B. Hotz, "Two Congressional Probes tackle Air Safety Problem," *Aviation News* (27 January 1947): 7–8; "TWA Ready to Use GCA-ILS," *Aviation News* (27 January 1947): 28; Austin Stevens, "Gander Radar Test Guides Planes In," *New York Times* (10 January 1947): 7.

21. T. P. Wright to James C. Johnson, 19 November 1945, file 818.1: ASR vol. 1, box 285, CAA Central Files, RG 237, NA.

22. CAA's personnel are estimated in Committee on Interstate and Foreign Commerce, *Safety in Air Navigation*, 80th Cong., 1st sess., January 1947, 244. The cut was not CAA specific. Congress had passed the FY 1946 budget and then passed the Ramspeck Act, raising the pay of federal civil service employees 14 percent percent. It then refused to go back and fix the payroll budgets, leading to widespread furloughs in all federal civilian agencies. See ibid., 237; and Wilson, *Turbulence Aloft*, chap. 6.

23. Assistant Administrator for Federal Airways to Administrator, 21 February 1947, file 818.1: ASR vol. 2, box 286, CAA Central Files, RG 237, NA. Unit maintenance files in RD 1276–1280, Records of Wright Field (Sarah Clarke Collection), RG 342, NA.

24. John Stuart, "Aviation Accidents: Ground Control Approach System, Backed by Many, Runs into Cost Objections," *New York Times* (2 February 1947): B-14. De Florez had submitted Alvarez's name to the Collier committee, and, as a member, had spoken in support of his nomination.

25. *AOPA Pilot* (November 1945): 42b–42c.

26. Brig. Gen. Alden Crawford to Commanding General, Air Material Command, 24

July 1946, file 413.44: AN/MPN-1, box 1870, AAF Unclassified Central Decimal Files, RG 18, NA.

27. The three sets used highly modified experimental displays that required only two operators, one for the search set and one for the two precision sets.

28. Alonzo Hamby, *Man of the People: A Life of Harry S. Truman* (Cambridge: Oxford University Press, 1995): 218, 248–60.

29. Anthony Leviero, "Congress Aroused over Air Crashes," *New York Times* (19 January 1947): B-16.

30. Ibid.; "Pressure Builds Up for GCA," *Aviation News* (6 January 1945): 3.

31. C. B. Allen, "Radar-'ILS' Rift Studied by CAA Chief," *New York Herald Tribune* (24 November 1946): 45; William Kroger, "Collier Trophy Award to Alvarez Spotlights GCA Development," *Aviation News* (16 December 1946): 7–8.

32. Committee on Interstate and Foreign Commerce, *Safety in Air Navigation*, 572–73 (see n. 22).

33. Allen, "Radar-'ILS' Rift," 45; Kroger, "Collier Trophy," 8.

34. T. P. Wright to Bob Sibley (Aviation Editor, *Boston Traveler*), 24 January 1947, file 818.1: ASR vol. 2, box 286, CAA Subject File, RG 237, NA. This letter is based on one provided to Wright by the Asst. Administrator for Federal Airways, 10 January 1947, file 504.0: vol. 11, box 84, ibid. Stanton held the office at that time, and the memo was actually written by C. M. Lample, Director of the Air Navigation Facilities Service.

35. Wright to Sibley, 24 January 1947. Emphasis in original.

36. Director, Air Navigation Facilities Service to Asst. Administrator for Federal Airways, 10 January 1947, file 504.0, vol. 11, box 286, CAA Subject File, RG 237, NA.

37. "AOPA Cites Needs for GCA Units," *AOPA Pilot* (January 1946): 42a.

38. Committee on Interstate and Foreign Commerce, *Safety in Air Navigation*, 1431, 1436, 1434.

39. Ibid., 1431.

40. Ibid., 1451.

41. Ibid., 570–71. He used "route miles" for his argument.

42. George E. Hopkins, *The Airline Pilots: A Study in Elite Unionization* (Cambridge, MA: Harvard University Press, 1971); George E. Hopkins, *Flying the Line: The First Half Century of the Air Line Pilots Association* (Washington, DC: ALPA, 1982). Behncke was a United Airlines pilot until his union presidency became a full-time job.

43. David Behncke, "Technically Speaking," *The Air Line Pilot* (March 1946): 5.

44. David Behncke, "Airlines Pilots Stake in Safety," *Air Transport* (June 1947): 27–30.

45. ALPA does not have locals. Instead, its regional councils have members from each represented airline that operates in that region; David Behncke, "Pilots Prefer ILS by 46–1 Margin," *The Air Line Pilot* (November 1947): 3, 5.

46. Nick A. Komons, *The Third Man: A History of the Airline Crew Complement Controversy, 1947–1981* (Washington, DC: Federal Aviation Administration, 1987); "New Radar Tests by Army Air Force," *New York Times* (15 February 1947): 8.

47. Committee on Interstate and Foreign Commerce, *Safety in Air Navigation*, 351.

48. Ibid., 489, 804.

49. Ibid., 805.

50. Ibid., 21; "Text of Air Safety Report," *New York Times* (3 July 1947): 8; House Committee on Interstate and Foreign Commerce, Subcommittee on Air Safety, report prepared by Carl Hinshaw, *Aids to Air Navigation and Landing*, 80th Cong., 1st sess., 11 July 1947, 17–18.

CHAPTER 8: TRANSFORMATIONS

1. "Text of Air Safety Report, *New York Times* (3 July 1947): 8.

2. Ibid.

3. U.S. Strategic Bombing Survey, Military Analysis Division, *Weather Factors in Combat Bombardment Operations in the European Theater*, 2nd ed. (January 1947), 9A. The difference in the two organizations' minimum ceiling requirements was due to Bomber Command's policy of night bombing, which demanded a higher ceiling in the absence of an effective low approach aid. The Eighth Air Force bombed, and landed, by day and hence could afford a lower ceiling within the same technological constraints.

4. David B. Langmuir to Lee DuBridge, 14 January 1943, file: Project 102, box 404, RD 158, Records of Wright Field (Sarah Clarke Collection), RG 342, NA.

5. Slightly more than a hundred collisions between bombers occurred in the thousand or so flying days the Eighth Bomber Command existed. About half occurred over England. See Roger A. Freeman, with Alan Crouchman and Vic Maslen, *Mighty Eighth War Diary* (New York: Jane's, 1981), 19.

6. George Comstock, 2 May 1945, file 1022 GCA, box 19, MIT Radiation Laboratory Records, RG 227, NA. See various correspondence in file: GCA correspondence [1945], ibid.; Acting Technical Assistant to Director, ANF Operations Service, 17 June 1946, both in file 818.1: vol. 1, box 285, CAA Subject File, RG 237, NA.

7. Glen Gilbert to Chief, Technical Development Division, 14 December 1945, file 818.1: ASR vol. 1, box 285, CAA Central Files, RG 237, NA.

8. Chief, Air Traffic Control Division to Chief, Technical Development Division, 14 September 1945; Air Navigation Facility Operation Service, "Operational Tests in the Use of Ground Radar Aids to Instrument Landing and Air Traffic Control," 1 June 1945, both in file 818.1: ASR vol. 1, box 285, CAA Central Files, RG 237, NA.

9. T. P. Wright to Mr. [Donald M.] Stuart, 30 November 1945; J. H. Miles to Donald M. Stuart, 30 April 1946, both in ibid.

10. Acting Technical Assistant to Director Air Navigation Facilities Operations Service, 17 June 1946; T. P. Wright to Asst. Secretary of Commerce, 26 December 1946, both in ibid.

11. Committee on Interstate and Foreign Commerce, *Safety in Air Navigation*, 80th Cong., 1st sess., January 1947, 768.

12. Edwin E Aldrin, "Fair to Foul Weather Flying," *Aeronautical Engineering Review* 5, no. 7 (1946): 26–29; "Civilian Application of Radar Moves Forward on Two Fronts," *Aviation News* (11 February 1946): 12; Deputy and Assistant Chief of Bureau of Aeronautics to Assistant Secretary of the Navy for Air, 24 January 1947, file: Equipment, Materials, Supplies, 1947, box 19, Bureau of Aeronautics Arcata Landing Aids Experimental Station Records, RG 72, NA.

13. The one-a-minute rate was a CAA estimate. The AAF believed that only one aircraft every three minutes could be accommodated, due to deflection of the ILS beams by the leading aircraft.

14. Arthur C. Clarke, *Glide Path* (New York: Harcourt, Brace & World, 1963), chaps. 23, 24. Milton Arnold claimed that there had been four FIDO installations in England by the end of the war, and that FIDO did not stand for anything—it was just a code word; "FIDO for Los Angeles," *Aero Digest* (April 1949): 60.

15. The average figure for Los Angeles was 2.5 percent, using a 200-foot ceiling. I have found no data presenting an average for all U.S. airports, or even the ten busiest. For comparison, La Guardia's average was 1.7 percent, again using a 200-foot ceiling. From Edgar A. Post, "Airline Operating Experiences with the Instrument Landing System," *Institute of Aeronautical Sciences,* 22 July 1949.

16. David L. Behncke, "Invitation to Progress," *The Air Line Pilot* (February 1946): 2. See also "Technically Speaking," *The Air Line Pilot* (March 1946): 5.

17. John M. R. Wilson, *Turbulence Aloft: The Civil Aeronautics Administration amid Wars and Rumors of Wars, 1938–1953* (Washington, DC: FAA, 1979), 237–41.

18. CAA coined these terms to eliminate confusion over what "GCA" meant. "GCA" was used in the media to variously represent all radar, the actual AN/MPN-1 equipment, all landing procedures controlled from the ground regardless of equipment, and just the short-range precision radars. By splitting the AN/MPN-1's three radars into two categories, ASR and PAR, CAA hoped to clarify its policies and improve aviation writers' understanding of its policies.

19. E. M. Sturhahn to Administrator [Wright], 30 January 1947, file 818.1: ASR vol. 2, box 286, CAA Central Files, RG 237, NA.

20. Wilson, *Turbulence Aloft,* 225–28.

21. Ibid.

22. Com. Paul Goldsborough, RTCA Executive Committee Meeting minutes, 6 April 1944, binder 1944-1945-1946, courtesy RTCA, Washington, DC. ARINC had been established by the airlines during the early 1930s to produce aircraft radios, since the airlines felt they were not being well-served by existing radio companies. Rentzel replaced Wright as Administrator of Civil Aeronautics in 1948. L. M. Sherer, RTCA Executive Committee Meeting minutes, 7 December 1945, binder 1944-1945-1946, courtesy RTCA.

23. RTCA, "Recommended United States Policy," Air Navigation-Communication-Traffic Control, 28 August 1946; Wilson, *Turbulence Aloft,* 233; The "Buck Rogers future" comment is from Ben Stern to the Editor, *Business Week* (2 October 1945), file 818.1: ASR vol. 1, box 285, CAA Central Files, RG 237, NA.

24. Walter McDougall, . . . *the Heavens and the Earth: A Political History of the Space Age* (New York: Basic Books, 1985), 229–31; Federal Aviation Agency, *FAA Statistical Handbook of Aviation* (Washington, DC, 1959), 22.

25. Erik M. Conway, "Echoes in the Grand Canyon: Public Catastrophes and Technologies of Control in American Aviation," *History and Technology* 20, no. 2 (June 2004): 115–34.

CONCLUSION

1. This sketch is based on the opening statement of Gregory A. Feith, the Investigator-in-Charge, to the public hearing on the accident held March 24, 1998. See also the final investigation report: National Transportation Safety Board, "Aircraft Accident Report: Controlled Flight into Terrain, Korean Air Flight 801, 6 August 1997," NTSB/AAR 0001, 13 January 2000, ix.

2. Karl F. Kettler to the editor, *Aviation Week and Space Technology* (May 4, 1998): 6.

3. Roger G. Miller, *To Save a City: The Berlin Airlift, 1948–1949* (Washington, DC: Air Force History and Museums Program, 1998).

4. Langdon Winner, *Autonomous Technology: Technics-out-of-Control as a Theme in Political Thought* (Cambridge: MIT Press, 1977), 326. Emphasis added.

5. Brian Balogh, *Chain Reaction: Expert Debate and Public Participation in American Commercial Nuclear Power, 1945–1975* (New York: Cambridge University Press, 1991.)

6. Susanne K. Schmidt and Raymond Werle, *Coordinating Technology: Studies in the International Standardization of Telecommunications* (Cambridge, MA: MIT Press, 1998), 20–21.

7. See Richard Hallion, *Legacy of Flight* (Seattle: University of Washington Press, 1976).

8. On SAGE, see Kent C. Redmond and Thomas M. Smith, *From Whirlwind to MITRE: The R&D Story of the SAGE Air Defense Computer* (Cambridge, MA: MIT Press, 2000). For Project Beacon's interest in SAGE as an air traffic control system, see Richard J. Kent, *Safe, Separated, and Soaring: A History of Federal Civil Aviation Policy, 1961–1972* (Washington, DC: FAA, 1980), 29–34; and Erik M. Conway, "Echoes in the Grand Canyon: Public Catastrophes and Technologies of Control in American Aviation," *History and Technology* 20, no. 2 (June 2004): 115–34.

9. *Twelve O'Clock High*, 20th Century Fox, 1949. Based on a book by Lay Beirne, *Twelve O'Clock High* (New York: Harper, 1948).

10. See David DeVorkin, *Science with a Vengeance* (New York: Springer-Verlag, 1992), 3.

11. Dale O. Smith, *Screaming Eagle: Memoirs of a B-17 Group Commander* (Chapel Hill: Algonquin Books, 1990), 64.

12. Frank Griffiths, *Angel Visits: From Biplane to Jet* (London: Thomas Harnsworth Publishing, 1986), 98–115.

Index

DATE DUE

Demco, Inc. 38-293